TRIM
IDEA BOOK

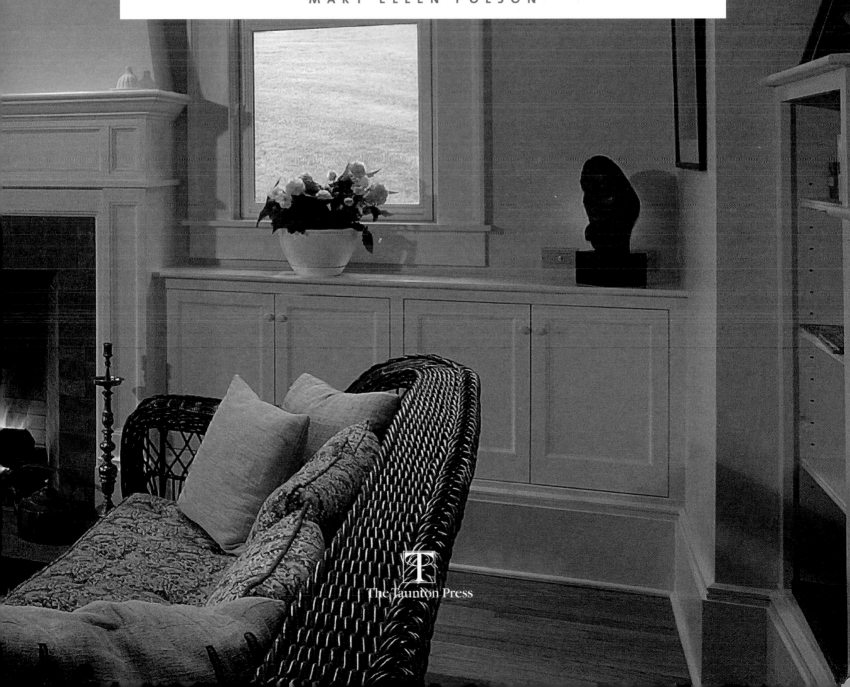

TRIM
IDEA BOOK

MARY ELLEN POLSON

The Taunton Press

The Taunton Press, Inc., 63 South Main Street, PO Box 5506, Newtown, CT 06470-5506
e-mail: tp@taunton.com

EDITOR: Jennifer Matlack
JACKET/COVER DESIGN: Jeannet Leendertse
INTERIOR DESIGN: Lori Wendin
LAYOUT: Laura Lind Design
ILLUSTRATOR: Christine Erikson
COVER PHOTOGRAPHERS: Front cover (top row, left to right): Roe A. Osborn, © The Taunton Press, Inc.;
© Chipper Hatter; © Tim Street-Porter; Charles Bickford, © The Taunton Press, Inc.; (middle row, left
to right): © Tim Street-Porter; Rob Karosis/www.robkarosis.com; © Brian Vanden Brink, photographer;
© Robert Perron, photographer; (bottom row, left to right): © Brian Vanden Brink, photographer;
© Brian Vanden Brink, photographer; Scott Gibson, © The Taunton Press, Inc.; © Chipper Hatter.
Back cover (clockwise from top left): © Brian Vanden Brink, photographer; © Tim Street-Porter;
© Scot Zimmerman; © Brian Vanden Brink, photographer.

Library of Congress Cataloging-in-Publication Data

Polson, Mary Ellen.
  Trim idea book / Mary Ellen Polson.
     p. cm.
  ISBN 1-56158-710-9
  1. Trim carpentry. I. Title.
  TH5695.P65 2005
  694'.6--dc22
                                    2004018188

Printed in the United States of America
10 9 8 7 6 5 4 3 2

Acknowledgments

Most writers have dreams of writing a best-seller. I have one in the works too, an in-depth biography (or perhaps a fictionalized novel, or maybe a children's book) about someone famous who died in the 1920s. It will come as no surprise to those who know me that I managed to finish this book long before I finished the first chapter of my alleged best-seller. Nothing motivates a writer more than a real deadline, or the pleasure of working for people who will actually publish your work.

Many thanks go to the editors at The Taunton Press who patiently worked with me and encouraged me throughout this process, especially Maureen Graney, Carolyn Mandarano, Jen Matlack, and Marilyn Zelinsky-Syarto. Thanks, too, to Sam Peterson, who suggested my name to Maureen when she was searching for new writers.

I also owe a world of thanks to my longtime editor and friend at *Old-House Interiors*, Patty Poore, for graciously allowing me to work on this book. Just about everything I know about decorative trim is a result of my experience at Gloucester Publishers, where interior ornament is as much a fact of life to us as the air we breathe. Everyone there deserves thanks for their friendship and support: Inga Soderberg, Bill O'Donnell, Becky Bernie, Lori Viator, Sharlene Gomes, Joanne Christopher, Grace Giambanco, and Julia Hite. Thanks also to my former colleagues at *Old-House Journal*, Gordon Bock and Josh Garskof, who taught me the importance of clarity and accuracy in writing.

Thanks, too, to my good friend Catherine Lundie, a fine writer herself, who is teaching me that I can enjoy my writing—and life itself—more.

Last but not least, I want to thank my husband, Jim Polson, whose influence and support go back more than 25 years. A superior writer and journalist since the day I met him, he is the real reason I call myself a writer today.

Contents

Introduction

I have to admit I'm a sucker for trim. The first house I ever bought—a 1940s Cape—had simple, slightly curving, one-piece trim where the wall met the ceiling. To me (raised in a succession of Ranch boxes in the '60s and '70s), such fine detail seemed the height of luxury. Although I didn't know it at the time, that simple bit of trim enhanced the classic proportions of each of the four small rooms, the dimensions of which were all divisible by three (9 x 9, 12 x 15, 12 x 18).

No wonder I fell in love when I saw the crown molding in the dining room of my next home, a 1923 Colonial Revival. No doubt the original trim had been identical in appearance to the molding in my little Cape, but the seller had the foresight to replace and double up the trim. The effect—achieved with stock pieces from the local home center—made the crown molding look far deeper and grander than the original had been, especially in the dramatically painted room. The effect of the rich, dark blueberry paint against the crisp white trim effectively sold us the house.

Years later, it finally occurred to me that if the owner of a house built with trim could pump up the volume, so to speak, why couldn't homeowners whose houses lack trim do the same? Indeed, they can, and in so many ways that we needed an entire book to show you just a few of the most dazzling examples. Not all trim is traditional, and not every trim material is wood or plaster. No matter how young or old your home, no matter what your personal style, trim can make your rooms look better, and more important, *feel* better. That's the motivation behind the *Trim Idea Book*. Use it with abandon.

Trim Basics

The ideal home is blessed with abundant space, kissed with light, and versatile enough to accommodate lifestyles that range from families with children to empty-nesters to single adults. In the house of your dreams, the rooms are spacious and adaptable: Bedrooms multitask as offices or dens; open-plan kitchens function as living or family rooms.

But whether you moved in yesterday or five years ago, you may struggle with turning your dream into the real thing. Spaces may seem out of proportion, cold, or dull, no matter how many times you rearrange the furniture or paint the room. Perhaps the element to transform your home hasn't occurred to you: trim.

Whether you call it woodwork, architectural molding, or plasterwork, room finishers like casings around doors and windows—and room "furniture" like wall paneling or fireplace mantels—are often the key to bringing out the best in a space, providing warmth and interest, and making it fully three-dimensional.

TRIM 101

Traditional trim is a form of decorative ornament that comes from classical architecture. The simplest types of trim to understand are moldings—linear pieces of wood or hardened plaster that you can find at your local home center or builder's supply store.

◀ A COLONNADE—a room divider composed of half walls and half columns—maintains a sense of openness between two rooms. The rooms are clearly separate, yet flow easily together, an important characteristic in houses with small rooms.

▶ IN A ROOM WITH FLAT WALLS and crisp corners, a crown molding where the wall meets the ceiling adds a decorative touch. The blocky, square shapes called dentils play a starring role here, especially when seen from a distance.

▼ A VAULTED CEILING adds a lot of drama to a room, but without some sort of architectural detail, the effect can be overwhelming. Wall moldings, window casings, and additional trim molding on and above the fireplace help anchor this vaulted bedroom.

A piece of molding reads as a straight line, but the face of the trim is usually a combination of recessed and projecting curves. Once trim molding is installed on a wall or the ceiling, the trim casts subtle shadows that help give adjacent flat surfaces dimension. The effect can be dramatic, from making a small room seem cozy by adding it around a fireplace opening, for example, to imposing a sense of proportion on rooms of unusual shapes. In most cases, trim applied in the right places will make a room feel more comfortable.

The reason has to do with our sense of depth perception. If you've ever tried to focus a camera on a blank wall, you know it's almost impossible—the lens has no point of reference. Turn the camera toward a chair rail or crown molding on the same wall, and voilá, the camera "sees" the wall and brings it into focus.

The human eye works in much the same way. Trim, with its undulations and curves, gives the eye a place to focus. It increases our perception of a room's size and depth, almost on a subliminal level. That's the reason trim can make a room seem larger.

Not all trim elements are the narrow strips of wood, plaster, or synthetics called moldings, of course. One of the oldest types of trim is the architectural column, which can support a ceiling, create a doorway or passage when used in pairs, or even create the presence of an open wall when aligned in rows. Since large pieces of house "furniture"—stairs, fireplaces, built-ins, and the like—tend to become focal points, it's also a good idea to think of them as elements in the overall decoration of a room.

DRESSING UP WITH TRIM

The walls and ceilings in many homes today are finished with drywall, a material that creates absolutely flat walls with crisp corners and edges. Painted off-white, the effect is clean and bright, but perfectly smooth white walls can also seem glacial. Add trim where the wall meets the ceiling—a molding that contains a row of the small, blocky shapes called dentils, for example—and those flat, boring walls take on a look of elegance. Trim around doors and windows— better known as casing—tends to be almost an afterthought, but it tells our senses that a portal is at hand. Casings also frame a view: out the window, through a door, or from one room into the next.

▼ WE HARDLY NOTICE CASINGS—the trim around a door opening—but they act much like picture frames to surround a view. Here, a wide door opening frames a classically proportioned fireplace, seen through a second, almost identical doorway.

◄ THE LINES IN THE PANELING and the trim around the door and windows in this foyer create a feeling of depth, which makes the room appear larger than it would if all of the surfaces were flat.

smaller, it will look as if it has more depth and dimension. In large spaces, you can be even more dramatic. Finishing a high ceiling with wood boards, for example, will warm up a voluminous room and give the space a definite shape.

Trim Transformations

Trim transforms spaces. It can be decorative or simply frame a view. Or it can perform such magical tasks as creating a sense of balance in an oddly shaped room—even bringing a vaulted ceiling back from orbit. Trim can enhance small and large spaces, conceal flaws, or define and separate spaces that act as individual rooms. It also can create unity in a single room or adjacent ones. In the form of house furniture like stairs and built-ins, trim can even be a functional part of the house.

While we all love the sense of space we get from a vaulted ceiling, there's often a disconnect between the functional part of the room—the living space—and the faraway ceiling. By adding a horizontal run of trim on the wall about 8 ft. above the floor, you can make the room feel cozier without sacrificing the sense of openness.

Used effectively, trim also can make small, undistinguished spaces seem special. For example, dress up a foyer by paneling it chest-high with beadboard—a type of paneling with a pattern of interlocking, vertical slats that comes in 4-ft. by 8-ft. sheets. Rather than making the room appear

► IN LARGE, OPEN-PLAN HOUSES, a pitched ceiling can cover a vast amount of territory. Paneling the ceiling with wood gives it a strong connection to the wood trim elsewhere in the house and gives shape to the enormous space.

The Difference Trim Makes

ROOMS THAT LACK TRIM often have a flat, two-dimensional quality that makes them seem sterile. Windows may resemble gaping holes rather than portals of light, and doors may look raw and unfinished. A lack of trim can be especially deadly in rooms that are on the small side, as well as in rooms that are overly large.

Simply adding the most basic types of trim—a baseboard at the juncture between the floor and the wall, trim casings around the perimeter of doors and windows—adds punch to the space and gives the eye somewhere to focus. Adding more trim can make an even greater impact. Think of it as set dressing. Just as a series of backdrops on a Broadway stage can trick the eye into believing that a landscape goes on forever, wall paneling or a fireplace mantel creates a greater sense of depth in a room. The result is a room that's more interesting, and seems larger than it actually is.

ROOM WITH MINIMAL TRIM

ROOM WITH ENHANCED TRIM

▶ USED AS A POST between a kitchen counter and the ceiling, a railroad tie is structural, but its real function is to signal the divide between the open kitchen and the adjacent living room without sacrificing a sense of free-flowing space.

▼ TRADITIONAL TYPES OF TRIM make flat surfaces look and feel more three-dimensional. Although the linear pieces of trim called moldings are often straight, they're shaped with curves—an idea repeated over and over in this oval room.

Trim Defines Spaces

One of the best aspects of large types of architectural trim—especially beams and columns—is that they can define and separate spaces. A rustic beam on the ceiling between a kitchen island and the family room signals the edge of the two rooms even when there are no walls present. A construction called a colonnade—two partial walls topped by columns on either side of a wide opening—is part doorway and part wall. It creates a passage between one room and the next and at the same time, delineates different areas. Neither beams nor columns interfere with the flow between rooms or the sense of openness. Rather, they enhance the space they occupy.

Even when walls are present, trim helps smooth the transition. Door casings, in particular, can unify adjacent rooms with different wall colors or purposes. As long as the trim is consistent, the rooms will flow together naturally. More obviously, using different kinds of trim such as crown molding and window casing in a closely related style is a good way to bring unity to a room with a lot of trim—a kitchen in a Victorian Revival style, for example. In minimalist settings, use trim to repeat an element already in the room to enhance its drama.

◄ LONG, NARROW HALLWAYS can be space-wasters in many homes. Here, the living room is separated from the hall with trim in the form of a half wall topped with a pillar, a sort of tapered column.

Trim Conceals Flaws

Just like people, rooms have flaws that we'd like to conceal. Let's say your living room is too angular: It has a high, vaulted ceiling with a pitch that ends sharply at the top of a wall. Paint the walls a bold color, and the room may make you feel claustrophobic. Leave the walls white, and you're left with a soaring space that points into a corner.

To steer attention away from the sharp ceiling pitch, create a focal point on one of the end walls. Paneling, built-in cabinetry, or even paint will serve to anchor the wall. The result is a room that has both focus and contrast.

▲ THE VICTORIANS WERE FAMOUS for their lavish use of trim, especially millwork. This Victorian Revival kitchen gets its unified look by combining and repeating many trim elements typical of the era, painted in a coat of crisp, clean white.

▶ WHITE WOODWORK unobtrusively separates rooms of dramatically different colors, allowing each space to create its own mood. At the same time, the rooms still connect to each other because they share the same trim details.

▲ MOLDINGS CAN CREATE A FOCAL POINT even in modern, minimalist settings. Here, horizontal bands of trim on the walls pick up and repeat the horizontal pattern of the stair railing, turning the entire stair area into a focal point.

▲ IN A LIVING ROOM WITH A SHARPLY PITCHED CEILING, full-height tropical-wood paneling adds weight and warmth to the angular lines of the space. The off-center positioning around the natural focal point of the fireplace helps bring the entire room into balance.

Trim Lives Large

While the term trim can be applied to a piece of molding only ½ in. thick, it also can be applied to large built-in elements many feet high, like staircases, fireplace mantels, and cabinetry. Rather than think of such house furniture as separate from the room itself, integrate these pieces into your overall design.

In the following pages, you'll see many ideas that will help you decorate your home with trim, and hopefully help you solve problems that keep it from being the beautiful dwelling you envision. Don't miss the resources in the back—that's where you'll find the tools you need to turn your dream home into reality.

▲ A STAIRCASE is one of the largest elements in any house, and it's almost entirely made of trim. Think of its design as part of an overall decorative scheme for your house, and you'll have a beautiful focal point.

◄ REPEATING THE TRIM ELEMENTS found in another part of the house is a great way to make a kitchen addition fit in. The vertical slat-and-spindle details in the island and the overhead divider were copied from a staircase in the original part of this house.

Relief for Walls

Many of us grew up in homes with small, boxy rooms where the walls met the floor and ceiling at sharp right angles. The only relief came from looking out the window or at a bright or soothing wall color. Today, our homes are often grander with more interesting features like vaulted ceilings and two-story foyers. But even these spacious rooms can feel cold or unfinished. Just as your yard may need a little land-scaping, your rooms could benefit from a few architectural trim elements to bring them into scale and make them feel warm and welcoming.

Baseboards, chair rails, paneling, crown moldings, and carved enrichments are readily available forms of architectural detailing that anyone can use to transform a room. These pieces and strips of shaped wood or plaster not only serve as transition pieces from one area to another, but they finish a room visually, adding variation to walls or ceilings that would otherwise seem monotonous. Even if you don't think of yourself as a traditionalist, just about any material with relief, from rough-hewn boards to textured wallpaper to pressed tin, can be used to trim a room.

◄ HALF WALLS TOPPED WITH TAPERED PILLARS are a good way to separate rooms with different purposes without sacrificing precious space and a sense of openness. By repeating details from one room to another, the two spaces are linked together.

From Baseboard to Crown

A T ITS SIMPLEST, A WALL CAN BE FINISHED with baseboard at the bottom. The area above the baseboard can be left plain or decorated in any number of ways, from the addition of a wainscot (a form of wall paneling) to a crown where the wall meets the ceiling.

Each of these elements, from the plainest band of baseboard to the most elaborate crown, can be built up from individual pieces of trim, commonly called moldings. Moldings are shaped strips of wood, plaster, or synthetic materials that create the appearance of lines and curves on flat surfaces. The more curves and lines in the molding, the richer and deeper the profile of the trim appears.

▲ TRIM ELEMENTS, INCLUDING CROWN MOLDING, raised-panel and beadboard wainscoting, door trim, and an acorn-shaped newel post help convert what is basically a series of rectangles—doorways seen through doorways—into a long, restful view.

◄ MANY OF THE MOLDINGS WE THINK OF AS TRADITIONAL come from period styles of the 18th and 19th centuries, such as Georgian and Federal. Well-proportioned trim in this new dining room includes paneled wainscoting, deep crown molding, and the fanlight over the double doorway.

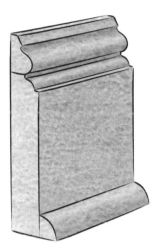

BASEBOARD
Baseboard is trim, often in the form of a flat strip of wood that covers and protects the lowest part of a wall where it meets the floor.

QUARTER-ROUND
Quarter-round is a rounded piece of trim with a profile that is nearly one-quarter of a circle, quarter round often finishes off the bottom of the baseboard.

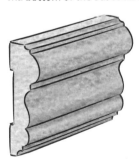

CHAIR RAIL
Chair rail is a horizontal strip of trim installed on a wall about 3 ft. above the floor. Functional as well as decorative, chair rail protects the wall from abrasion by the backs of chairs, and often tops off paneled wainscoting.

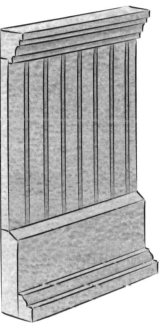

WAINSCOT
Wainscot is a decorative or protective facing, such as wood boards, on the lower portion of an interior wall.

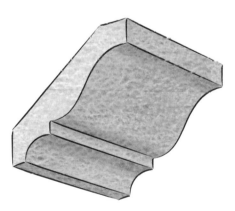

CROWN MOLDING
Crown molding is the trim at the very top of an interior wall where it meets the ceiling.

TRIM IN SMALL SPACES

▶ WHILE A PIECE OF CROWN MOLDING reads as a straight line, it's actually a combination of recessed and projecting curves. That's why it harmonizes with curves elsewhere in this foyer, such as the serpentine staircase and the curving edge of the lowest stair step.

▼ TRIM CAN ADD A SENSE OF DIMENSIONALITY that helps a small room seem larger. Relatively simple elements like a beadboard wainscot and simple casings around doorways and windows help define and separate living and kitchen areas into two distinct spaces.

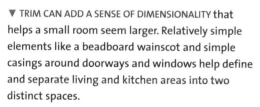

▶ EXPOSED WALL STUDS ADD DIMENSION to a room and make it seem larger than it actually is. In this small dining area, shallow strips of wood called battens applied to the walls are used to similar effect. A distressed ceiling beam complements the rustic table.

▲ REMOVING A LOW CEILING and exposing a network of joists, beams, and rafters transforms a small bedroom from a cramped, stuffy space to a relaxed, airy one. The exposed beams effectively become the trimwork for the room.

Molding vs. Millwork

MOLDING AND MILLWORK are not one and the same. Moldings are a linear form of decorative ornament that comes from classical architecture. Used primarily on interior surfaces, moldings can be cut or carved from wood, or molded from plaster or synthetic materials. Millwork, on the other hand, is a product of the Industrial Age. It's the term for house components and related trim that are sawn from wood, such as doors, casings, porch posts, brackets, and window sashes.

TRIM IN LARGE SPACES

▲ EVEN THOUGH THIS LARGE CONTEMPORARY LIVING ROOM has a vaulted ceiling, the builder didn't leave out the crown molding. Because it has relief and detail, the minimalist band around the perimeter of the ceiling still works like traditional molding to bring the room into scale.

▶ AN OPEN STAIR with a wrought-iron railing might seem spare and cold in this broad hallway if the surrounding walls weren't finished with deep crown moldings. The deep moldings over the arched doorways at the front and back of the hall reinforce the warming effect.

◀ CLASSICAL TRIM ELEMENTS such as columns, crown molding, a chair rail and wainscoting, and an elegant stair rail work together to bring a large space into graceful proportion. Despite its complexity, the room is both interesting and restful.

Creating a Traditional Baseboard

IF YOU CAN'T FIND A BASEBOARD you like at your local home center, design your own by combining two or more stock pieces of wood. First, find a flat or low-relief trim board that measures 4 in. to 6 in. high and install it along the bottom of the wall, tight to the floor. Next, add a piece of trim molding with a rounded (convex) profile, such as a basic quarter-round, at the bottom of the flat or low-relief trim board, where it will cover the edge of the floor. Finally, to create a smooth transition between the top of the baseboard and the wall, cap the edge of the flat or low-relief board with a recessed (concave) profile. The entire assembly should be between 6 in. and 8 in. high.

▲ ALTHOUGH THE PARTS for this elaborate Victorian baseboard have been custom-milled, you can approximate a similar look by breaking the profile down into components, then shopping for stock trim that closely matches the chosen model.

▲ LARGE ROOMS WITH LOW CEILINGS can seem monotonous, even oppressive, without architectural detailing. This living room gets welcome relief due to the massive dark beams that span the ceiling and the floor-to-ceiling wall paneling, painted a crisp white.

COMMON CROWN MOLDING PROFILES

It is possible to create the crown molding profile you want by combining stock or custom elements.

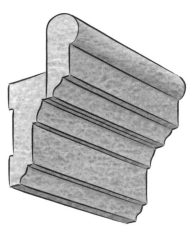

EXAMPLE 1
Two-piece crown molding with a relatively flat profile

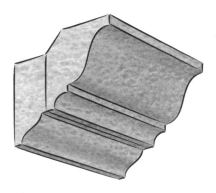

EXAMPLE 2
Two-piece crown molding with deep, shallow profile

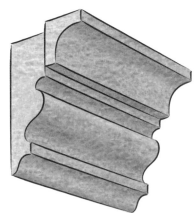

EXAMPLE 3
Two-piece crown molding with convex top piece and elaborate back piece

▼ ENORMOUS BY CONVENTIONAL STANDARDS, this great room still feels comfortable because it includes trim elements that bring the room into scale. The white trim over the entry, fireplace, and windows pulls the eye upward, breaking up the space. The trim also highlights the light wall color.

▲ TO SHOW OFF THE DEPTH of a coved ceiling crown, the owner of this 1913 Philadelphia home painted both the walls and the ceiling a warm coffee brown. Other white wall accents help balance the darker color.

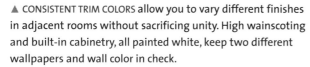 CONSISTENT TRIM COLORS allow you to vary different finishes in adjacent rooms without sacrificing unity. High wainscoting and built-in cabinetry, all painted white, keep two different wallpapers and wall color in check.

▶ FOR ROOMS PANELED with beadboard or knotty pine, it's traditional to finish the trim with the same treatment as the walls to enhance the woodsy feel. The unpainted pine walls and ceilings in this summer camp are naturally colorful because they darkened as they aged.

▲ ALMOST ANY COLOR can work in a high-ceilinged room, especially one with trim. Abundant light pouring through the windows keeps the bright yellow walls from looking dark and dingy, while moldings and casings keep the bold color from becoming overpowering.

Wall Colors and Finishes

THE OLD DECORATING TRUISM "WHITE UNITES" makes perfect sense when it comes to choosing trim colors. Even if you paint the living room teal and the dining room salmon, you can achieve a neat, unified look if the trim framing the doors and walls of the two rooms is the same shade of white.

But not every wall is painted, of course. If your walls and trim are both wood, for example, they have a natural unity that allows you to match the two elements perfectly, or to heighten the sense of contrast between them. Bleaching or whitewashing wood leaves the walls almost white; stains can turn even a light wood to dark walnut. Pigmented stains add a sense of color while allowing the grain to shine through. For a subtle effect, play lighter or darker trim against medium-toned wood walls. For a bolder contrast, make the trim dramatically darker or lighter than the surrounding walls, or color it with stain or paint.

▲ EXPANSES OF IVORY-WHITE WOODWORK, including the ceiling, wainscoting, and door and window casings, complement and balance vivid rose-pink walls in this dining room. Against a softer color, expanses of white trim would give a more subtle look.

▼ DON'T OVERLOOK the possibilities of rustic materials for trim. Exposed beams and rafters may be part of the structure of a house, but they also can create a sense of pattern and order in a dining room with an exposed ceiling and walls.

▲ HISTORICALLY, ELABORATE MOLDINGS higher up on the wall were often cast of plaster, a medium that lends itself to fine detail. While skilled plaster artisans are hard to find outside the Northeast, you can find plaster components at home centers or specialty stores.

Traditional Trim Materials

TRADITIONAL MOLDINGS are made out of plaster, synthetic materials, or wood. While plaster—a wet material that cures quickly after it's shaped—is still the standard for ornament, installation requires a degree of craftsmanship that most homeowners don't have. Ready-to-install moldings are sold through some home centers and the Internet.

Plaster alternatives, including both rigid and flexible ure-thanes and other synthetics, can be shaped into any form that plaster takes on, but they must be painted or finished, or they'll look like plastic. Wood is the most versatile and durable medium for trim because it's relatively inexpensive and can be cut into almost any shape.

▲ IN NEW CONSTRUCTION, it's often more cost-effective to create detailed moldings, like the dentil pattern in this fireplace mantel, with cast synthetic materials like urethanes and polymers. Once painted, these lightweight materials are indistinguishable from wood or plaster.

◄ ANOTHER TRADITIONAL TRIM MATERIAL is wood, which is relatively inexpensive to shape, buy, and install. It's preferred for door and window casings, as well as anything likely to be bumped or scuffed, such as chair rails, baseboards, and wall paneling.

Wainscots and Paneling

WHILE WE AUTOMATICALLY THINK OF PAINT AND WALLPAPER as typical decorative finishes for walls, paneling takes the idea to a new level. Paneling can appear at almost any part of a wall: at the bottom, at the top, and from floor to ceiling. Wainscoting usually covers the lower portion of a wall. Except for the flat plywood paneling you may remember from your basement rec room, even the simplest beadboard paneling adds to a room, giving it a subtle three-dimensional effect.

Although wood is the traditional material for paneling, almost any textured material can be used to create wainscoting, including paint, patterned or textured wallpaper, textured plaster, fabric, leather, or pressed metal. Depending on the finish treatment, paneling can brighten a room (whites and neutrals), deepen its mood (dark varnished woodwork), bring nature indoors (unfinished fir or teak), or create a sense of great age (deep, saturated colors).

▲ TO CREATE A NATURAL LOOK that also provides some relief for bare white walls, let the wood speak for itself. This wainscoting is a composition of flat pine boards framed top and bottom with horizontal bands of the same species.

◄ VERTICAL KNOTTY-PINE PANELING creates its own mood, especially as it mellows with age. The wood's tendency toward reddish-yellow hues as it matures makes it the perfect foil for cottage-style furnishings that include whites and blues.

▲ SHALLOW STRIPS OF WOOD, called battens, and small brackets supporting the plate rail add relief to the high wainscoting in this dining room. Had the wainscot been merely a flat expanse of white, it would have been too stark against the dark green wall.

▲ THE ARCHED WINDOW OVER THE BED and light-colored wainscoting help ground this bedroom with multiple sloped ceilings. Oddly shaped ceilings tend to make even large rooms look smaller, so trim helps to outline and anchor the usable space.

▶ WALNUT-STAINED, ARTS AND CRAFTS– STYLE WAINSCOTING enfolds an entry hall. High, dark wainscoting gives a sense of being surrounded by a warm enclosure. This type of trim is traditional in dining rooms, foyers, and libraries, where a little bit of formality goes a long way.

Chair Rails and Plate Rails

WELL-PROPORTIONED WAINSCOTING can be high or low, but it never falls exactly halfway up the wall, where it could visually split the room in two. As a rule of thumb, think in thirds: Most wainscoting is either chair-rail height, reaching about one-third of the way up the wall, or plate rail height (two-thirds of the wall height). For a room with 8-ft.- to 9-ft.-high ceilings, chair-rail height wainscoting will usually fall between 32 in. to 36 in. up the wall. Higher wainscoting, like that typical of Craftsman dining rooms, will reach 60 in. or more. In a room with higher or lower ceilings, you do the math: Divide the height of the room by three, and snap a chalk line at a height that visually pleases you.

CHAIR RAIL
A chair rail's profile is usually shallow, with a more prominent lip near the top to keep chairs from striking the wall.

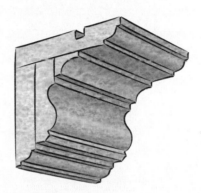

PLATE RAIL
A plate rail has a deep profile. It is made up of a narrow shelf or rail mounted about two-thirds of the way up a wall, with a horizontal groove down the middle to support plates.

▲ IN THIS FARMHOUSE PARLOR, the ribbon-like wallpaper pattern is a subtle variation on the vertical lines in the white beadboard wainscoting, which reaches just high enough to give the room a sense of balance.

PICTURE MOLDING
A picture molding, or rail, has a shallow profile since it is a thin piece of trim placed around the perimeter of a room 1 ft. to 2 ft. below the ceiling, with a slot or curved lip at its top. The lip allows just enough space to hang a picture hook behind the molding.

Batten Paneling

YOU CAN PANEL A WALL inexpensively with battens, narrow boards called furring strips that are about 2½ in. to 3 in. wide by ⅜ in. deep. Apply the strips vertically at regular intervals along a wall to create batten paneling. Finish it at the top with a slightly wider, 3-in. to 4-in. horizontal strip, with or without a plate rail. You can paint between the battens, apply textured or flat wallpaper, or apply panels of a rough-textured fabric like linen or burlap. Paint the fabric or textured wallpaper in the color of your choice.

▲ WAINSCOTING PAINTED OR FINISHED in rich colors adds a sense of dimension to any room—even utilitarian ones like these adjacent entry halls. To maintain visual continuity, keep wainscoting at the same height in adjoining rooms.

◀ THE SMALL ATTACHED COLUMN, known as a colonette, supporting the mantelpiece is clearly one of the focal points of this living room. In the background, low, olive green wainscoting accents the colonette's unusual shape without calling attention to itself.

▼ TRIMMED WITH A BRIGHT YELLOW CHAIR RAIL, the buff-white wainscoting is a splendid excuse for color in this cheerful living room. The color has been carefully chosen to enhance the owner's collection of artwork. The colorful trim and molding also help to frame the canvases.

FLOOR TO CEILING PANELING

▲ CROSS-BATTEN PANELING serves multiple purposes in a compact sleeping area. Visually, the horizontal and vertical lines of the batten extend outward, making the space seem larger. From a functional standpoint, the paneled wall serves both as a divider and as a built-in headboard.

▲ REPEATING PANELING ELEMENTS is a classic method of creating balance and proportion in a room. Here, one side of the room is a mirror image of the other: two doors, two flat columns called pilasters, even two swags, one on either side of the mantel.

WALL PANEL PARTS

One of the most traditional patterns for wall paneling is the raised-panel style. Flat, rectangular panels seem to float between stiles (vertical strips) and rails (horizontal strips) in a repeating pattern. In this simple interpretation, the beveled edges make the face of the panels appear to project out from the stiles and rails.

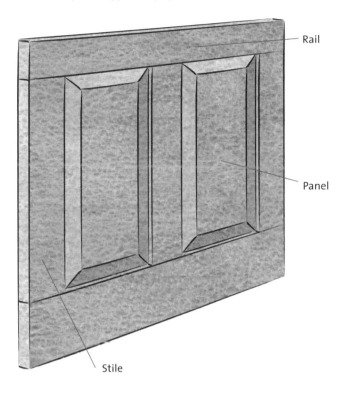

Rail

Panel

Stile

◄ PANELING MADE FROM LOCAL WOODS— in this case, California redwood— creates its own ambiance by regional association. Here, wide, horizontal boards serve as a rich, dark counterpoint to the open, almost weightless lines of the adjacent glazed doors. Cutouts in the top panels add interest.

▲ THE WIDE GROOVES between the rough, horizontal boards on both walls and ceiling add interest and help give this low-eaved bedroom some dimension. Other trim details, including an arched door and a mantelpiece supported by brackets, introduce softening curves.

▶ CRISP AND FRESH, new knotty pine, fir, or alder paneling is much lighter in appearance than varnished wood that's had time to age. Knots and other natural features in the wood add character to this simple, rustic paneling.

◀ THE REPEATING LINES of a hallway paneled with wide boards prepare the way into a more formal room with chair-rail-high wainscoting. Even though the spaces are very different, the common element of paneling creates a smooth transition from one room to the next.

▼ LAUAN WOOD PANELS outlined with narrow battens create a dramatic grid pattern on a dining room wall. Large, rectangular shapes work especially well with high ceilings because they provide some visual.

COLORS AND FINISHES

▶ BOARD PANELING can be distinctly contemporary. In this tall, narrow foyer, horizontal boards add texture and color. Using vertical boards would make the high-ceilinged area feel small and claustrophobic. The light color of the paneling also helps to open up the space.

◀ THE EXPOSED JOISTS in this camp-style bedroom have mellowed over the years, taking on the same patina as the bare-board walls. The romance of old places like this are one reason new homes are built with exposed walls today.

◄ TO HIGHLIGHT THE WARM HUE of the small dotted line on the wall stencil, the raised-panel wainscoting is stained a matching shade. Typically, wainscoting along a staircase runs upward on the diagonal, and the panels are parallelograms, not rectangles.

Coloring and Finishing Trim

LTHOUGH WHITE has been one of the most enduring finish colors for trim, trim finishes can be far more versatile, lending themselves to almost any interior color scheme. For example, when paints were in short supply in colonial times, color was usually applied to the trim rather than the walls. Favorite shades included deep, earthy browns, blue-greens, and reds, or creamier yellows and blues achieved with the addition of milk solids, which created milk paint.

Later, when millwork became more common, varnishes allowed the natural beauty of the wood to shine through. Today, you can paint trim any color under the rainbow, including white, tint it with pigmented stains, or stain it darker or lighter. In the past, many traditional varnishes and shellacs allowed wood to gradually darken. Modern water- and oil-based polyurethanes contain preservatives that allow you to maintain your color choice indefinitely.

▲ A DEEP, BRICK-RED COLOR on woodwork and wainscoting in an Early American gathering room reflects the colonial preference for painting trim rather than walls. The unpainted walls provide relief from the intensity of the casings and paneling.

Enrichments

ENRICHMENTS ARE THE CRÈME DE LA CRÈME of interior ornament. Think of these purely decorative elements as you would icing on a cake: pretty to look at, but not essential to every room. An example of a simple enrichment would be a plain piece of trim molding applied to a wall, such as a chair rail. A more elaborate enrichment might be a recessed shell with flutings and carvings, installed over a corner cupboard.

Applied moldings and architectural carvings have their origins in classical and Renaissance architecture and decoration, where they were just as likely to appear on vases or furniture as on walls and ceilings. The term enrichments covers a wide range of decorative products applied to walls and ceilings, from traditional plaster and wood carvings to lightweight synthetic materials.

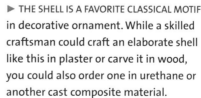

◄ CHAIR RAILS, ONE OF THE SIMPLEST FORMS of wall moldings, not only break up the monotony of plain, unadorned walls, but they also protect the surface from damage from repeated contact with furniture. This chair rail unifies a room with two wall colors.

► THE SHELL IS A FAVORITE CLASSICAL MOTIF in decorative ornament. While a skilled craftsman could craft an elaborate shell like this in plaster or carve it in wood, you could also order one in urethane or another cast composite material.

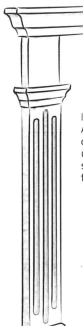

▲ PANEL MOLDING may resemble an empty picture frame, but used skillfully, it can create order, proportion, and beauty in a room. In this elegant sitting room, panel molding of different shapes is used in groupings to center large elements like a recessed niche and an arched doorway.

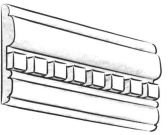

BRACKET
A projecting element, usually in the form of a horizontal scroll or block

DENTIL
A series of small, square, tooth-like blocks, typically across the band of molding in crowns

FRIEZE
A horizontal band near the top of an interior wall immediately below the crown that may be plain, or decorated with wallpaper or enrichments

PILASTER
An almost flat, two-dimensional column with some sort of relief, often fluting

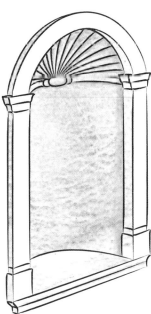

NICHE
A recess in a wall, often in the shape of an arch or semi-circle

ENRICHMENT VARIETIES

▲ ONE OF MOST POPULAR TYPES of wall enrichment is the pilaster, a sort of flat column that is often fluted, or in this case, reeded. Pilasters are a classic way to frame door openings or divide wall spaces into regular intervals.

◄ PANEL MOLDING ADDS INTEREST to a room painted white. Under a beamed ceiling, this high wainscoting is enriched with simple vertical panels. A single horizontal panel molding over the mantel is a subtle way to unite the fireplace to both walls and ceiling.

About Enrichment Materials

CONTRARY TO POPULAR BELIEF, plaster is much lighter than wood—and so are many plaster and carved wood synthetic look-alikes, making it easier to work with these materials for larger installations. Cast plaster ornaments such as brackets and niches can be shipped directly from specialty dealers to your home. Wood-carved ornaments can usually be found in home centers and on the Internet.

Larger enrichments are more likely made of a composite material, which is less fragile to ship. These plaster and wood look-alikes are usually made from flexible polyester resins, urethanes, or gypsum reinforced with fiberglass. The best are durable, non-flammable, and indistinguishable from fine plaster or ornaments carved from wood, provided they're painted or given a decorative finish. Composite materials usually come from specialty retailers, although you may find some on the Internet.

▲ WORRIED ABOUT PAINTING an elaborate enrichment once it's installed? Real plaster is naturally white and doesn't necessarily need paint. For both plaster and synthetic enrichments, you can use water-based (latex) paint, and even use a spray-painting tool to get the paint into crevices.

Victorian Gingerbread

I N THE LATE 19TH CENTURY, trimwork milled from wood was so abundant and affordable that builders used elaborate embellishments known as gingerbread everywhere, giving Victorian homes their distinctive look. On the outside, houses were ornamented from top to bottom, from cut-and-scrolled decorative boards called bargeboards in the gable to latticework underneath the porch. Inside, open doorways and halls were favorite places for decorative pieces of cutwork known as spandrels. Ornamented with scrolls, spindles, medallions, lattices, and scalloped brackets, many traditional Victorian patterns can be re-created easily today by specialty millworks.

▲ THE SCROLLING BRACKETS AND BEADED SPINDLES in a fanciful Victorian spandrel enliven molding over the sink in a new kitchen. The gingerbread isn't just for show: It also shields task lighting from view, eliminating glare.

◄ WHITE TRIMWORK AT THE BASEBOARD, chair rail, around windows and doors—even the paneled doors in a built-in—all work as wall decoration in a Wedgwood-blue room. The white accents help establish some continuity in a room with unusual bump-outs under the tray ceiling.

► THE AVAILABILITY OF MOLDING and other forms of trimwork in flexible synthetics like urethane makes it easy to install baseboard along an undulating wall. The traditional method of curving baseboard or other wood molding requires making cuts along the back of the piece at intervals.

Accent on Ceilings

Think of your ceiling as a waiting canvas. While there's nothing wrong with painting it with one or two coats of parchment white, finishing a ceiling with trim can add drama to a room, make it seem more or less formal, and even correct imbalances of proportion.

The materials for ceiling trim range from traditional to contemporary and refined to rustic. For a formal, dressed-up look, finish the edges of the ceiling with crown molding and add enrichments such as applied moldings or a medallion. For a more casual effect, trim the ceiling with wood paneling or add a recessed tray ceiling.

If you want to add a finishing touch to an especially large room or to one with a vaulted ceiling, consider beams. Beamed ceilings can be as elegant as finished paneling, or as casual as rough-hewn timbers.

No matter which trim effect you choose, consider color part of the overall scheme. Paint moldings a soft white or emphasize them with subtle colors; darken or lighten wood beams or even paint or tint them to achieve the effect you're after.

◄ INSTEAD OF LEAVING the exposed walls, joists, and rafters natural, the owners of this summer house chose to paint them white, giving the space a cooler effect. Although the room is still cozy, it easily adapts to more formal furnishings.

Crown Moldings Revisited

ROWN MOLDING DOES MORE THAN COVER THE GAP where the wall meets the ceiling. Even the plainest piece of ceiling trim makes a room less box-like. Just as modest crown molding can make a small room seem larger by increasing our sense of depth perception, deep crown molding in a room with high ceilings will make the room seem warmer and more human in scale.

There are a number of ways to vary crown molding. A favorite classical treatment is cove molding, where concave- or bowl-shaped trim makes it appear that the wall is actually curving toward the ceiling. Another option is the tray ceiling, a type of dropped ceiling with a recessed center. Tray ceilings are a practical method of decorating a large, high-ceilinged room without much trim. They are also a dramatic way to recapture some of the space under a gable roof without using a sloped ceiling.

▲ EVEN OUTDOOR ROOMS like this large, screened porch can be topped with crown molding. In this case, the high-ceilinged room also has a frieze—the band of white trimmed with vertical battens painted a pleasant green.

▶ AN EMBOSSED CROWN MOLDING in this mint green bedroom gives the appearance that it's actually lifting the ceiling, which makes the room seem more spacious. The crown molding is pressed metal, or tin—a popular trim material with a lot of potential for ceiling decoration.

▲ CROWN MOLDINGS are among the most elegant and graceful types of trim; even deep ones can be surprisingly simple. The key to success is achieving the right balance of curving and flat surfaces in proportion to one another.

▲ ONE WAY TO GIVE KITCHEN CABINETRY a custom, built-in look is to install deep crown molding over a run of wall cabinets. You can even disguise a soffit—a box built to conceal an exhaust or heating vent—with trim.

Creating Crown Moldings

YOU CAN CREATE a crown molding with as little as two pieces of stock trim. Deeper, more elaborate styles require a minimum of three pieces. The best way to make a crown you like is to experiment with short pieces of stock from a builders' supply store or home center.

For a three-piece crown, you'll need two relatively flat pieces of trim about 4 in. to 6 in. high by 1 in. thick. Apply one piece along the wall and the other against the ceiling. Both pieces should have a concave curve or recessed right angle on one end. The third piece, the crown proper, installs over the two backing pieces at an angle that allows some of both the flat and curving edges of the backer trim to show. The crown piece should have concave and convex curves on its face for best effect. It should also have at least one section that's flat and parallel to the wall when the trim is in place to provide a greater sense of depth.

For an even deeper crown, use blocks of wood behind your trim pieces. These backing pieces will allow you to create a more complex molding.

▲ A DEEP CROWN MOLDING gives the impression that it's a single piece of solid wood with multiple profiles. In reality, even skilled carpenters create crown moldings with a combination of stock and custom-cut trim pieces.

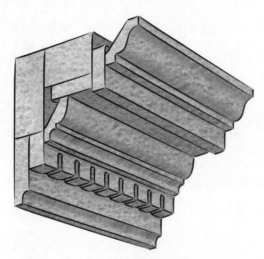

Multiple-piece crown with backer blocking

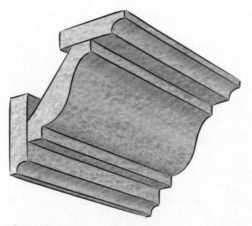

Three-piece crown

◀ EXTENDING THE TRIM over a pair of matching doors up to the ceiling crown molding creates extra shadow lines and adds depth to the overall application. The play of light and shadow across the varied surfaces of a crown molding is what makes it so appealing.

▼ ADDING PICTURE MOLDING approximately 2 ft. below the crown molding in a room with an especially high ceiling has the effect of making the crown appear larger. Here, a wider, deeper crown lowers the perceived height of the ceiling without sacrificing the room's sense of spaciousness.

◀ PROJECTING DENTILS—a row of blocks in the crown molding— make this formal, period-style room seem three-dimensional. They complement other high-relief accents in the room, including the mantel columns and the keystone, the central element at the top of an arch above the cabinet.

▲ THE DROPPED CROWN MOLDING creates both warmth and elegance in this high-ceilinged room. Paired with rich gray walls, the bracketed crown molding serves as a stylish yet cozy enclosure for the pencil post bed. The beams add a feeling of security and ground the room.

Cove Ceilings

COVE MOLDINGS CREATE THE ILLUSION that the wall and the ceiling meet in a smoothly arching curve. Today's wood and synthetic cove moldings often have a gap at the back, making these elements an ideal way to conceal wiring or, in the case of a dropped cove molding, recessed lighting. Like crown moldings, a cove molding can be installed with two pieces of trim (see "Creating Crown Moldings" on p. 52). For the crown piece at the center of the cove, choose trim molding with a deeply concave curve.

▶ THE COVE MOLDING in this Victorian-era bedroom extends from the picture rail on the wall to the panel molding on the ceiling. It is relatively easy for a skilled craftsman to cove a ceiling using plaster before applying other trim.

▼ A COVE MOLDING'S SMOOTH, curving surfaces lends itself to decoration, but avoid painting it in high-contrast colors. As a rule of thumb, apply darker paint on the recessed areas, alternating with lighter paint on the more prominent trim. This will enhance the dimensional effect.

▶ YOU CAN CREATE THE LOOK of a coved ceiling with a piece of concave trim in wood or a synthetic material, provided it's deep enough, and then finish the bottom with simple trim molding. Here, the cove molding above the fireplace complements the shape of the mantel.

TRAY CEILINGS

▶ A TRAY CEILING offers more than one surface for decoration. In this formal dining room, the recessed part of the tray gets a deep crown treatment, while the lower part of the tray is edged with flat trim.

▼UNDER A SHALLOW VAULTED CEILING, shelf-like extensions trimmed with crown molding create a tray effect on two sides of a bedroom. Hidden behind the crown, recessed accent lighting plays up the gentle curve of the ceiling.

▶ THINK OF A TRAY CEILING as a wide, overhead picture frame. The shallow tray in this bedroom accentuates the room's minimalist, modern lines. In color and shape, the ceiling is almost a mirror image of the low platform bed.

Lighting Up Tray Ceilings

TRAY CEILINGS ARE GREAT PLACES TO INSTALL LIGHTING. **With the right equipment, you can achieve almost any effect, from soft ambient light to task and accent lighting.** For an ambient effect, install narrow-diameter fluorescent or pencil-sized low-voltage halogen strip lighting in recessed areas. The light will bounce off the part of the ceiling above the tray and reflect soft, diffused light downward. For task or accent lighting, mount small recessed or puck-shaped halogen disk lights on the bottom of the tray. These versatile, low-voltage downlights can illuminate a counter or a piece of furniture, or swivel sideways to highlight artwork.

▲ RECESSED DOWNLIGHTS mounted on the face of a tray ceiling create their own decorative pattern and do double duty as accent or task lighting. Create ambient lighting by adding low-profile light strips in the recess over the tray.

Applied Ceilings

▲ PLASTER MEDALLIONS are a time-honored way to decorate a formal room or to add a touch of grandeur to a simpler one. This lightweight medallion cast from polyurethane is surprisingly easy to install; it can be painted or even gilded.

SOME CEILINGS DON'T NEED ANY MORE DECORATION than a simple edging of crown molding. Others, though, can benefit from more. In rooms that have a lot of trim, a related ceiling application can help unify the space. Rooms that lack decoration can look finished and sophisticated if the trim goes on the ceiling rather than the walls.

Just about any decorative material can be applied to a ceiling, from glow-in-the-dark stars to strips of peeled bark. You can't go wrong with standard materials like plaster and wood, however. If you only want to add a sense of texture and color to the room, consider an almost flat material, like board paneling. To give the ceiling a more three-dimensional effect, add a formal treatment like rope molding, or choose a type of paneling with some relief, such as a coffered ceiling.

◄ THE SPARE LINES of modern rooms can sometimes seem cold or sterile. This dining area avoids that fate because the board-and-batten ceiling adds both color and texture while accentuating the lines of other elements in the room.

▲ ON A SHALLOW BARREL-VAULTED CEILING, dark double battens are a natural continuation of the trim elsewhere in the room, from the window casing to the wood-capped pillar and rail at the head of the stairs.

PLASTER

▶ ENRICHMENTS AREN'T FOR EVERYONE, but this classically inspired ceiling shows a wide range of possibilities. Appliqués—thin strips of flexible material that form decorations like swags and beaded moldings—are applied separately, like pieces in a complex jigsaw puzzle.

▲ LIKE A TRAY CEILING, applying panel molding to a ceiling creates a subtle picture frame for a room. This ribbon-like example re-emphasizes the room's rectangular shape and is close enough to the crown molding that it adds a greater illusion of depth.

◀ FINE PLASTER ORNAMENT demonstrates superior craftsmanship and deep, three-dimensional relief. While synthetic versions are usually cast as one piece, the decorations on a medallion are often cast separately and then individually glued to the base.

▲ ARCHED STRIPS OF TRIM applied at regular intervals along a narrow hallway emphasize the beautiful shape of a barrel-vaulted ceiling. The trim also has the effect of making the hall seem longer than it actually is.

▶ PLASTER ORNAMENT FOR CEILINGS, or appliqués, are often delicate in appearance, like the panel molding surrounding the medallion in this double parlor. Contemporary man-made moldings are made of flexible, easy-to-apply polyester resins, which are prefinished and come ready to paint.

WOOD

◄ THE WOOD-SHEATHED VAULTED CEILING in this large living room shares a color kinship with the red tile floor. In large rooms with little trim, choosing complementary materials for the floor and ceiling is a good way to unify the space.

▲ THE OWNER of this 1912 Prairie-style house found a clever way to conceal the presence of two cross beams on his dining room ceiling. He applied strips of dark stained oak in an interlocking pattern that incorporates the beams into the design.

◄ IN THIS ROOM FILLED with metal and glass, the ceiling paneled with narrow wood strips has a grounding effect. Like a hardwood floor, the paneling introduces both warm color and a natural material without drawing much attention to itself.

Coffered Ceilings

COFFERED CEILING is one of the fanciest types of paneled ceilings. The sunken panels (or coffers) have at least four and as many as eight sides. These ornamental ceilings have been made of wood, stucco, plaster, and stone for centuries. Although they look rich and heavy, appearances can be misleading: The Romans coffered the ceiling of the Parthenon to reduce its weight.

If you like the look of these elaborate ceilings, prefabricated options might fit your wallet better than hiring an Italian artisan. A number of companies offer suspended, interlocking systems that resemble fine wall paneling in beautiful hardwoods like cherry and oak.

▲ COFFERING CREATES THE ILLUSION of great depth on a ceiling. The deepest part of the coffer is usually flush with the ceiling; it is recessed by building up trim molding on as many as eight sides.

Beams

AN EXPOSED BEAM SERVES MANY FUNCTIONS in a room—some are structural, some purely decorative. Others, whether by intention, inspiration, or pure accident, are both.

Envision a vacation cottage studded with exposed wall joists and rafters. The boards and beams are structural, but through long association with lazy afternoons at the shore, they also create a certain mood. In other words, they're decorative. Contemporary timber-frame dwellings take this idea one step further: The cross beams and ceiling joists not only hold the house in place, they are as much a part of the interior decor as any other kind of trim.

Obviously, not every home has timberwork that can do double duty as trim. But you can certainly add beams to create a unique look in your home. You'll be in good company; beams have been an intimate part of home design throughout the world for centuries.

▲ THESE CROSS BEAMS not only support the ceiling, but they're also the focal point of the room. In effect, they make a frame for the skylight. Painted wood paneling on the walls and ceiling accentuate the dark beams.

◄ EVEN THOUGH THESE TIMBERS appear to be structural, there is a good chance they're merely for show. In any case, the beams make a pleasing trim accent in rooms that would otherwise be all white.

▶ BOX-BEAM CEILINGS are another traditional treatment for formal rooms. The interlocking, crossing grid pattern makes the ceiling appear deep and substantial. Here, they're offset with floral wallpaper, which adds a dash of color to the room.

▼ THIS YEAR-ROUND CAMP-STYLE HOME in New Hampshire is planned around reclaimed timbers that support exposed joists throughout the house. The joists give the board ceiling (which probably hides insulation) the rustic look of a vacation cottage.

REFINED BEAMED CEILINGS

▶ BECAUSE THEY ARE USUALLY HOLLOW, beams are a good place to conceal wiring for lighting and electricity. The massive cross-beam in this kitchen, supported by a half column, also serves as a divider between the kitchen and adjoining room.

▲ BECAUSE THEY ARE THREE-DIMENSIONAL and extend outward from the ceiling, beams play much the same role as crown molding in a formal room. In this case, a large beam subtly divides the bedroom into sleeping and sitting areas.

▼ A BOX-BEAM CEILING is a form of false timbering, similar to the half-timbered effects on the faces of Tudor-style houses. The beams in this sitting room are fairly flat, but they look as though they are partly submerged in the ceiling.

Box-Beam Ceilings

BOX-BEAM CEILINGS are a classic way of giving a room a formal, well-proportioned appearance. Although box beams look like massive timbers finished with trim molding, the beams are actually made by fastening three pieces of board together to form a U-shape. The addition of trim molding makes the beams seem deeper and solid when in fact, they're hollow. With that in mind, box beams are a good place to conceal supports that cross the breadth of the room, such as joists or steel girders. They are also an ideal way to hide wiring and electrical boxes that accompany hanging light fixtures.

▲ A CLOSE LOOK AT A BOX BEAM reveals that it is an application, not a structural part of the ceiling. The cross-hatch pattern is decorative in its own right, and the flat spaces in between beams lend themselves to stenciling, wallpaper, or textured finishes.

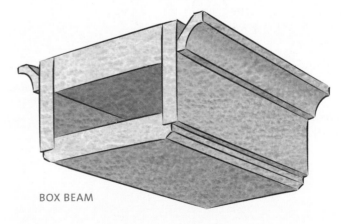

BOX BEAM

Adding Color Above

PLAY UP CEILING TRIM with a splash of color. With subtle applications like crown or panel molding, you're free to use color, but try to keep the shade of the molding lighter than the wall behind it. That's because the projecting trim will make the color appear more prominent, increasing the chances that the effect will be stark rather than understated.

Wood paneling naturally brings tones of brown, yellow, and red into a room. Accentuate the shades you like best with stain or a clear natural finish. Or you can bleach or color-tint the wood to deepen or lighten a color effect.

Beamed ceilings are among the most versatile canvases for color. The beams themselves can be painted, stenciled, bleached, darkened, or left natural. The same list of treatments can be applied to the ceiling proper, or area between the beams, plus at least one more: wallpaper.

Whatever combination of colors and textures you choose, make sure it's in keeping with the rest of your color scheme—and be sure that it isn't so dramatic that it overpowers the room.

▲ A NATURAL WOOD BOARD CEILING with exposed joists gets a subtle burst of color from light blue stain. In a play on the classic blue-white color combination, the walls have been pickled or bleached to turn them almost white.

◄ EVEN ELABORATE COLOR TREATMENTS can be subtle, if the tones and patterns are balanced. In this Flemish-style room, the stenciling on the beams relates to the intricate design on the ceiling panels and in the gable wall in both form and color.

▲ IF YOU CHOOSE A DARK COLOR for the beams in a room with matching paneling, paint the rest of the ceiling a lighter color to keep the room from appearing too dark. The lighter parts of the ceiling effectively become a neutral background.

◄ THE COLOR PALETTE in this dining room graduates from a pale blue-white on the beamed ceiling to deep blue on the wainscot. Painting the ceiling and the beams the same color has a calming effect on the room.

RUSTIC BEAMED CEILINGS

▼ VAULTED CEILINGS COVER a lot of real estate. Left undecorated, they can leave the eye with nowhere to focus—an experience that can be unsettling. The dark, rafter-like beams in this bedroom not only add drama, but they also bring the pitched ceiling down to earth.

◀ DARK, TRESTLE-LIKE BEAMS are traditional ceiling treatments in Southwest architecture. The widely spaced beams not only add texture to the soft, sun-saturated wall color, but they also accentuate other rustic woods in the room, from the mantelpiece to the furniture.

Using Reclaimed Beams

H ARVESTED FROM CENTURY-OLD FACTORIES and dilapidated barns, reclaimed beams and boards are an excellent way to introduce the look of aged wood into a new home. Much of the wood used more than 100 years ago was virgin, old-growth timber. Despite its old age, this tightly grained wood is still stronger than a lot of the wood being harvested today.

Many of these old boards and timbers carry the marks of earlier use, from the scars made by worms or insects to the scuffmarks of a hand-held crosscut-saw. These scars are greatly prized by builders and homeowners for the character they add to the wood. Incorporating salvaged wood into your home also means you'll prevent tons of material from reaching landfills and simply rotting away.

▲ SO ROUGH-LOOKING that the bark still seems to be clinging to them, these telephone-pole shaped beams somehow work in this sophisticated, eclectic bedroom. Tints in the beams accent a variety of colors and textures in the furniture and bed coverings.

▲ THE TIMBER FRAMING in this lake house getaway was resawn from old-growth Douglas fir reclaimed from demolished factories in the Northwest. If you are using a combination of old and new wood, buy wood of the same species and apply a uniform stain.

Creative Material Choices

Ceilings don't get a lot of wear and for that reason are great places to have fun with imaginative finishing materials. You can use everything from corrugated metal, gold leaf, or stencils, to fabrics, wallpapers, or birch bark to dress up a ceiling. One of the most versatile and popular types of applied ceilings is the pressed metal or "tin" ceiling. Relatively inexpensive and easy to install, pressed-metal ceilings are actually made of copper or steel rather than tin. Each interlocking 2-ft. by 4-ft. sheet is made up of multiple rectangles impressed with the same pattern. You can even finish your choice with edging and crown molding made from the same material.

▼ IN AN ATTIC BEDROOM, corrugated metal is an interesting choice for ceiling paneling, especially together with the exposed wood joists. Because the shiny metal is reflective, it brings more light into the room. The joists, window casings, and hardwood floor tie the overall space together.

▲ A PRESSED-METAL CEILING adds texture and shine to a contemporary kitchen. Tin ceilings, which are actually made of copper or steel, are also easy to paint; because the surface is embossed, the material lends itself to highlighting and polychroming (painting with multiple colors).

▲ BIRCH BARK is a popular finishing material for walls, ceilings, and furniture in woodsy settings like this camp-style house. The material is remarkably resilient, provided it's not subject to heavy use or exposed to the weather.

▶ AN ALL-WHITE SITTING ROOM under a gabled ceiling gains a sense of openness through careful exposure of just a few joists and rafters. The joists directly under the peak suggest a traditional horizontal ceiling, while open rafters on one side frame an angled skylight.

◀ TRADITIONALLY PAIRED with white stucco walls, dark, beamed ceilings are a favorite treatment in hot climates like the Southwest. Instead of feeling oppressive, the dark canopy actually suggests shade and protection from the sun.

Adornment for Doors

Think of trimming a door as a way to frame a view and you'll realize the possibilities for decoration are endless. The most basic trim is casing: flat or slightly curved, shallow boards mounted at the top and sides of a doorway. Casing doesn't just cover the gap where the wall ends and the door frame begins; it also announces the presence of the door, making it easier to locate. Like formal moldings for walls and ceilings, door casings can relate to popular styles or eras, such as Arts and Crafts or Victorian. Once you've chosen a look or style, it's a good idea to carry it continuously through your home.

Doorways can be arched, created by columns or posts or topped with elaborate molding elements like a classical frieze. The most practical and versatile material for door trim is wood. In cases where large vertical dividers like columns or posts actually create the door opening, the fine line between door and decoration disappears altogether: The column serves as trim and frames the passageway structurally. These broad passageways both unify and subtly divide different rooms, making both appear larger. At the same time, however, they clearly define each room as a separate space, much as a wall does.

◄ A SIMPLE, FLAT CASING emphasizes the presence of the doorway between living room and library. The strong rectangular lines of the door casing and the transom above it also play up the many rectangular trim elements in the room in a flattering way.

Door Casings

DOOR CASINGS ARE MADE UP of horizontal and vertical pieces of wood or other trim material. The top piece is the head casing. The vertical pieces are called side casings. While the side and head casings can meet in many ways, the seams at the corners should be almost invisible.

The simplest way to conceal edge lines is to butt the side casings to the bottom edge of the head casing. Although the 90° cuts are easy to make, it's hard to get the joints to disappear. In a mitered joint, the head and side casings meet on a diagonal at the corners. Skillfully cut, a mitered joint appears almost seamless. This method works especially well for casings with some relief. Other traditional decorative techniques include plain or carved corner blocks, banding around the outside edges of the casing, and top molding over the head casing.

▲ KEEPING CASINGS CONSISTENT will reward you when a door frames a view of another door or window. The casings on both of these openings have band edging on the inside and outside edges, giving them a sense of depth.

◀ WHEN A DOOR BARELY FITS under a low or sloping ceiling, fit the casing to the contours of the space to give it a cleaner appearance. The owner of this Early American house played up the unusual shape of the casing with dark blue-green paint.

▲ IN THIS ASIAN-INSPIRED BATH, the side casings on a set of closet doors meet the head casing on the perpendicular in an almost invisible butt joint. The head casing gets an extra flourish with the addition of crown molding.

CASING STYLES

▲ VICTORIAN-STYLE DOOR CASINGS often featured fluting or grooves cut into the wood, like those shown here. The side and head casings meet at a separate corner block, which can be plain or decorated with rosettes or other fanciful designs.

◄ ARTS AND CRAFTS–STYLE CASING begins with flat, plain boards, then adds bead molding and a simple crown. The bead between the vertical and horizontal casings is a good way to hide the seam between the pieces.

Style Options for Door Casings

IF YOUR DOOR CASINGS are plain and slightly curved, you probably have what's called clamshell casing. You can upgrade this inexpensive and neutral look by searching for stock components at your local home center.

One of the simplest ways to dress up flat 1-in. by 4-in. stock is by adding an edge of trim molding around the perimeter of the casing face. For a Victorian effect, install a plain or decorative rosette block at the top corners where the head and side casings meet. For Arts and Crafts styling, add a layer of thin, horizontal bead molding between the side and head casings. (The bead molding should project slightly over the side casings.) Then add crown molding over the top of the head casing. If you want something even more formal, use fluted stock casings for the casing sides to create a Neoclassical look.

▲ ALTHOUGH THE SIDE CASINGS on this doorway are flat and plain, the surround gains presence from decorative treatments on the head, or top horizontal, casing. Elements include bead and crown molding and a classical-style corner block.

TYPES OF CASING

Beaded

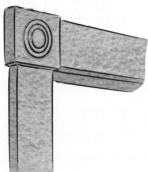

Rosette block

Arts and Crafts

Neoclassical

INFORMAL CASINGS

▶ BRIGHTEN DOOR CASINGS with color for a cheerful, informal look. In this newly constructed camp-style home, the owner outlined the doorways in a light turquoise green, then painted the kitchen cabinets a contrasting lemon yellow with just a hint of green to unify the colors.

Transoms

I N A SENSE, a transom is an extension of a head casing: It's a small window or panel directly over a door. Often transoms are hinged to open and close, allowing air to circulate from one room to another even when the door is shut. A useful means of brightening dark or interior rooms, transoms are also a good idea for settings where both cross-ventilation and privacy are desired, such as a bedroom or a bathroom off of a hallway.

▶ TRANSOMS BRIGHTEN dimly lit rooms. In a narrow hallway in an old house, this transom over the door brings in extra light. The mustard-yellow color applied to the casings and other trim brightens the interior and adds dimension to rough-textured plaster walls.

▲ THIS ENTRY IS FINISHED with weathered boards, probably reclaimed from an old barn or factory. Despite its rough-hewn appearance, there is a sense of formality: The slightly arched head casing is supported on either side by brackets.

◀ IN AN OPEN-PLAN HOUSE with rustic twig railings and tree-trunk columns, door openings are trimmed with the same knotty-pine wood used on the walls and ceilings. The trim works especially well on an oddly shaped door made of the same wood.

FORMAL CASINGS

▼ EVEN THE PLAINEST DOOR CASINGS can be dramatic when they're painted a strong color. Dark forest green and deep brick red are a classic combination; paint the door one color and the trim the other.

▼ SINCE THE MID-19TH CENTURY, millworks have been turning out casings that look hand carved but are actually machine made. While these were likely hand made, specialty retail stores and home centers can usually order look-alikes.

▲ THESE DOOR CASINGS are tapered, meaning the casing is thicker at the outer edge and thinner toward the door opening. Like a picture frame, recessing the casing inward tends to focus attention on the view beyond the door.

▲ ALTHOUGH THE CASINGS on this period door look hand sculpted, the fine detail was probably created with compo, an early type of composition material made from sawdust and glues. Today's pliable resin-based enrichments can be applied right over the casing.

Arches

AN ARCH IS A POWERFUL PRESENCE. Not only is the elegant, semi-circular curve of the arch a dramatic focal point, but it's also a particularly strong means of constructing an opening.

The curve of an arch can vary from deep to shallow, just like trim molding. Given that arches are symmetrical, they are an excellent choice for almost any size passageway. The many variations on the basic arch shape include bell, half-circle, and pointed.

Because the shape itself is ornamental, an arch can look finished without any trim at all. But you may want to add case moldings, decorative banding or wood trim, or a deeper molding finished with a decorative block at the top of the arch, called a keystone.

◄ ARCHES COME IN MANY SHAPES, styles, and sizes. The clean, contemporary lines of a wide passageway topped with a pedimented arch frames a more traditional bell-shaped arch in the room beyond it. The two work together because they're both variations on the same form.

◄ MUCH OF THE DECORATIVE IMPACT of this entry comes from the wood-plank doors that together form an arch. The curved glass panels inset into the doors reinforce the arch shape, while the white casing around the doorway blends into the walls.

► ARCHES DO A SUPERB JOB of framing a view into another room, especially when trim elements are classical and well-proportioned. The casing on this semi-circular arch is trimmed with a reed detail and finished with a decorative block, called a keystone, at the top.

▼ ARCHES CAN LITERALLY BE CARVED out of stone, sculpted from plaster, or in this case, cut from wood. Cutting the arch shape into the wall helps emphasize the rustic, rounded shape of the log paneling.

◄ ARCHES ARE PARTICULARLY BEAUTIFUL when they frame views of other arches. Here, a wide, smoothly finished interior arch transforms a wood-paneled entry door into a focal point. Rustic hardware and casings made of the same wood add an extra touch of romance.

Bending Casings

IN THESE DAYS OF FLEXIBLE DRYWALL, it's a fairly easy matter to create an open archway. Bending a casing to trim the arch, however, may prove more difficult. Kerfing, the traditional method for bending, involves cutting slots of uniform depth into the back side of wood that allow it to bend. The slots must be cut at evenly spaced intervals—if the cuts are made too close together the wood can crack or break apart.

Another technique is to steam-bend plywood molding or to build up layers of flexible plywood on a form that matches the shape of the arch. Obviously, these techniques are best left to skilled woodworkers and carpenters. Fortunately, it's also possible to buy flexible trim moldings that fit almost any radius or curve from specialty suppliers and some home centers.

▲ CREATE AN ARCHED entry by using multiple elements. Although this door has only the hint of an arch shape, flanking it with sidelights and adding a vertical-paned archway overhead make a dramatic statement.

▲ TRIMMING AN ARCH with case molding isn't as difficult as it might seem. In addition to traditional methods like kerfing, in which uniform slots are cut into the back of the casing to allow it to bend, specialty dealers also make flexible case moldings.

PLASTER

▶ BECAUSE ITS SHAPE IS rounded, an arch makes a fitting entry for this circular music room. Outlining the recessed area with dark paint helps the arch relate to the casings over the arched windows in the piano room.

▼ THE ARCHES in these interconnected hallways are unusual in that each spans the space from wall to wall. They are visually supported by projecting decorative blocks called corbels. Although corbels can be structural, these are purely decorative.

◄ A WIDE, OPEN ARCH is an elegant way to create a transition between a dining room and a living room. Arches that lack trim should be finished by someone with excellent drywall or plastering skills, since any flaws in the plasterwork will show.

▼ THE FORMAL CASINGS on these arched openings are finished with keystones, a decorative accent dating back to the invention of the arch by the Romans. Located at the center or topmost area of an arch, keystones are well suited for traditionally styled homes like a Colonial Revival.

About Keystones

THE EARLIEST ARCHES were made of brick or stone, built up toward the top center so that each stone or brick was supported by the preceding one. Once both sides of the arch were almost complete, a wedge-shaped gap remained at the top. The keystone—a slightly larger brick or stone cut to fill the gap—was born.

Keystones were a common sight over doorways and entryways in colonial and Revolutionary War times. For that reason, keystones continue to be popular in many housing styles with colonial roots. Today, placing a keystone in the top of an arch is a particularly easy task—when the arch is made of wood.

KEYSTONE

Columns

A COLUMN IS A LONG, VERTICAL SUPPORT, either cylindrical or rectangular, that is used to hold up a ceiling or a roof. On its own, a column is not really part of a wall, nor is it obviously part of a doorway. Yet pair two columns together and you've created a passage. Line three or more columns in a row and you have an open wall.

Columns can provide support—or they can give the appearance of doing so without being structural at all. These chameleons are composed of three main parts. The base supports the long part, called the shaft. The capital tops the shaft. Columns that rest on a high base, such as a half-wall (a built-in bookcase, for example), are called half-columns. While the column originated in classical architecture, these versatile structures can look traditional or contemporary, depending on the setting.

▲ A COLUMN IS AN ELEGANT WAY to support a beam or ceiling, especially when there are other classically inspired elements in a room. This smooth, cylindrical column with a simple Tuscan capital has a tapered shaft.

◀ IN SMALL SPACES, use half-columns over bookcases or other built-ins to help define a room without fully enclosing it. In effect, these short, tapered columns not only finish the open walls, they help create a passageway.

▲ RECTANGULAR WOOD POSTS do more than merely support an overhanging ceiling in this open-plan living area. Even though they're placed at least 12 ft. apart, they help to signal where one room ends and another begins.

COLUMNS AS DIVIDERS

▶ IN THIS LARGE, HIGH-CEILINGED ROOM, the owner has carved out a cozy, partially open nook for relaxing or web browsing. The columns not only complement the semi-circular shape of the half-wall, but they also allow light from the windows to reach the rest of the room.

▼ PAIRED TOGETHER, these rectangular posts support an overhead beam that serves as the dividing line between two large, open rooms. The posts signal that the area is really two rooms instead of one without actually closing off either space.

▶ TO ADD PRESENCE to a narrow end wall between a hallway and a living room, add a half-column resting on a paneled base. The other side of the base facing the living room could serve as a spot for a recessed bookshelf.

The Affordable Column

YLINDRICAL COLUMNS, which are usually tapered according to strict rules of proportion, are difficult and expensive to make using traditional methods. Fortunately, that has changed in recent years. It is now possible to buy a classically proportioned wood column for use in your home for as little as a few hundred dollars. These columns are usually called paint or stain grade, which means you'll need to finish them once they're installed. Cast columns made from reinforced polymers or fiberglass can be even less expensive. For example, an 8-ft.-high column that measures 12 in. across at the base may cost $300 or less. (The capital, the crown for the column, is usually extra.) Another option is cast stone, which you can find on the Internet. Many companies sell directly to homeowners and will work with you to design the pieces you need.

▲ A ROW OF TUSCAN COLUMNS divides a large living area into two distinct, yet open, spaces. The affordability of cast columns makes them an attractive option for homeowners who want to create a formal interior with an open floor plan.

COLUMNS AS PASSAGEWAYS

▶ LIKE BOX BEAMS, square columns are usually hollow. Used with low partial walls to create a formal entry into a large living area, these posts have been embellished with panel and crown molding—still formal but a different look than a cylindrical column.

◀ BRANCHING TREE TRUNKS create a novel entryway into a circular dining room with panoramic views. Although the look is highly decorative, the forking trunks also suggest great strength and support for a cross beam. The small forked tree above the dining table reinforces the theme.

The Capital

A CAPITAL IS A THREE-DIMENSIONAL CROWN for a column. The ancient Greeks and Romans who invented the column considered it to be an idealized version of the human form: The base represented the feet, the shaft the body, and the capital the head. The two most elaborate styles—Ionic and Corinthian—were considered feminine.

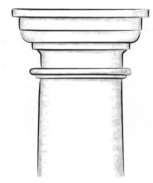

TUSCAN
Roll-shaped with a smooth column shaft

DORIC
Pillow or roll-shaped; closely resembles a Tuscan capital, but column shaft is fluted

IONIC
Four scrolling ram's horns, or volutes, at each corner with a fluted column shaft

CORINTHIAN
At least two flaring tiers of leaves; fluted and the tallest of the four types of columns

▶ THESE FLUTED COLUMNS are Ionic. While many classical styles of columns were fluted, Ionic columns were always topped by distinctive capitals with four curving scrolls. The design is thought to resemble a feminine hair style.

▼ ONE COLUMN in this open-plan kitchen is freestanding; the other is attached to a low half-wall. Together, they manage to create not one but two open walls, as well as two passageways into and out of the kitchen.

Window Enhancements

J ust as doors are portals to other rooms, a window is a portal to the outside. But while window casings are often similar in appearance to door casings, there are a few differences that make window trim unique.

The first is psychological. Where door casings subliminally alert us to the presence of a passageway, window casings make us aware that we're still inside, protected by walls and windows, even though we can see outdoors. Unlike doors, windows often occur in pairs or groups of three or more. The casings for these groupings can be kept separate, or combined for different effects, such as establishing a sense of style in a room, or conveying a feeling of formality or informality.

Windows usually have some sort of sill, or bottom ledge. It is often overlooked from the standpoint of decoration, and finishing it with trim molding or other detailing can add a sense of balance and proportion to the window frame. Since windows come in many shapes and sizes, choosing an oval, star, or other novelty shape with appropriate trim can really brighten up a room.

◄ THE PANORAMIC VIEW HERE would seem to demand a continuous picture window, but the contrast between indoors and out might overwhelm. The strong linear casings on the sashes (the two window parts that slide up and down) provide a sense of separation without impeding the view.

Window Casings

WINDOW CASINGS TEND TO REGISTER IN THE BACKGROUND of our sense of perception. For that reason, they are wonderful tools for establishing a theme or creating a tone for a room. If your goal is to create a certain style for your home, such as Cottage or Arts and Crafts, for instance, start with the window casings. If you get the details right, you'll have the bones of the look you want before you've purchased the first piece of furniture.

While the sash, the part of a window that slides up and down, and muntins, the dividers between panes of glass, aren't actually considered trim, their presence or absence has an impact on a window's overall appearance. From a design standpoint, it's hard to separate the casing from these window elements.

▲ WINDOW CASING not only serves as the finishing touch for a window, but it also adds extra relief to a wall. Painted a crisp white, the window casing and the baseboard serve as a counterpoint for the strong color on the walls.

▼ IN ROOMS WITH RUNS of uniformly sized windows, you can achieve a clean, unified look by running the casings together with banding at the top and a uniform sill at the bottom. This treatment works particularly well with small, narrow windows.

▲ THE WOODWORK in this large Craftsman-style kitchen doesn't overwhelm, in part because the casings over the doors and windows have been ganged together, resulting in a smooth, streamlined look. The natural warmth of the wood keeps the look cozy.

▲ SOMETIMES THE CASING NEEDS to play second fiddle to the design of the window. Painted the same color as the wall, this casing is subtle, simply outlining the elegant shape of an arched casement window with thin, horizontal banding.

FORMAL CASINGS

▲ ELEMENTS OF A FORMAL WINDOW CASING include a generously sized top (or head) casing, and deep sills. The repeating pattern of vertical and horizontal lines in this trio of casement windows in a dining area creates a strong rhythm, almost like a piece of artwork.

▶ TRIM FOR A WINDOW can go far beyond mere casings. In this circa 1830 Philadelphia home, for instance, there are gold-leafed rosettes at the corners, interior panel shutters inside the window frame, and paneling on the lower part of the wall below each window.

◄ CRAFTSMAN-STYLE CASINGS are often grouped together in twos and threes. Although the casings are usually flat, the top, or head, casing is usually topped with crown molding. As this bedroom shows, additional crown molding at ceiling height accentuates the casing details.

Divided-Light Windows

IN ARCHITECTURAL PARLANCE, a "light" in a window is a pane of glass. In earlier times, a window sash was built up from smaller individual panes of glass, separated by pieces of framing called muntins. While some companies still make true divided-light windows, most manufacturers usually create window sash from larger single panes of glass, then add the dividers separately.

The simplest grids simply snap on. While they give the appearance of a traditional divided-light window from a distance, up close you can tell they're fake—mostly because they don't throw shadow lines that create the appearance of depth.

A more sophisticated approach, called simulated divided lights, creates a more realistic appearance. In this case, the window is double glazed, meaning there are two parallel pieces of glass in each sash instead of one, with an air space between them. Muntin grids are then permanently affixed to each side of the window, and a third grid that shadows the external grids is sandwiched between the two panes of glass.

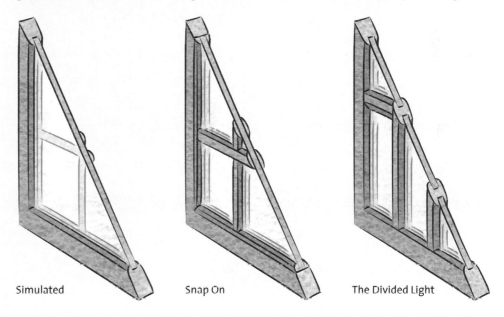

Simulated   Snap On   The Divided Light

◄ FLUTED AT THE TOP AND SIDES and finished with carved rosette blocks at the corners, Victorian window casings are similar to those on Victorian doors. One difference is the presence of a sill, which projects slightly at the bottom of the side casings.

INFORMAL CASINGS

▼ THEY MAY BE PLAIN AND SIMPLE, but these casings are highly graphic, outlining a T-shape in white on a green wall. Windows that slide open on the horizontal add an extra touch of whimsy.

▲ CASEMENT WINDOWS in an octagonal, ocean-view room are trimmed with finely grooved casings. Although the casings repeat the linear pattern of the beadboard walls, you don't really notice them—the eye is directed toward the view beyond the window.

◄ THESE NATURAL-WOOD CASINGS accentuate the Asian feel of the bathroom. The casings unite several windows of different sizes and also join with other woodwork in the room to give the space a clean, streamlined look.

▶ IN A DINING AREA with distressed beams on the ceiling and in the corner, the owner has opted to paint the window casings the same color as the walls. In effect, they disappear, allowing the rustic beams and the view to come to the foreground.

▼ SINCE THE VIEW ISN'T EXCEPTIONAL, the natural-wood casings become a focal point in this sunny dining area. The rich, red-brown color accents the apple-green walls. Slightly angling the head casings at the ends adds extra interest.

COLOR ON CASINGS

► A RUNNING LEAF PATTERN carved into the interior casing adds extra dimension to a berry-red window. It also sets the theme for other decoration in the room, including the pattern of the curtains and a stencil on the wall.

▼ IN A ROOM MOSTLY FILLED with books and woodwork, painting the window casings the same color as other trim in the room makes the space less busy. The cheerful red also brightens up the room.

▲ CLEAN WHITE CASINGS against a dark green wall allow a leaded glass window to shine over a combination bookcase-mantel. The casings are flat, but trim molding for the sill and bead molding below the top casing gives the window architectural presence.

▲ YOU DON'T EXPECT to find window casings at the top of a stairway landing leading to other rooms. To call attention to these casings from the room below, the homeowners painted them a bold color that contrasts with the light color of the walls.

◄ THE BRIGHT, SUNNY COLOR of this informal dining room plays up the Victorian rosette blocks at both the top and the bottom of the window casings. In particular, the crenelated detail over the top rosette blocks stands out in crisp relief.

◄ IT'S POSSIBLE TO HAVE a deep window sill without any casings at all. This window frame is set into the wall around casement windows. The deep sill is a handy place to display decorative items or grow plants.

The Window Sill

THINK OF A WINDOW SILL as a shallow shelf for a window. It can be as simple as a rounded 1-in. by 4-in. board that juts below the rail and casings, or it can be more elaborate. A typical finishing technique is to add an apron—a sort of baseboard for windows—below the sill. You can dress up the apron with trim molding applied just below the stool (the horizontal part of the sill). The apron also can be cut away or fluted at the sides and bottom to add extra relief.

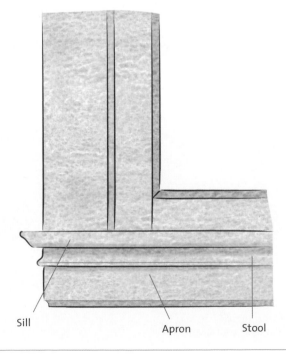

Sill   Apron   Stool

◄ ALTHOUGH THE WINDOW CASINGS in this kitchen have been stained to match the natural-wood boards, the sill is painted a deep blue. The same color is used on the cabinet fronts and dividers in the windows to unify the space.

▲ IN A GABLED ROOM with sloping ceilings and low side walls, a window sill with a deep apron anchors the space below a window. The light-colored apron unites with all the other trim in the room, helping to bring order to an oddly shaped space.

◄ A WINDOW SILL can be as simple as a board with a rounded edge inserted beneath the side casings. Here, the entire casing was painted the same mustard-yellow color as other trim, adding extra architectural dimension to rough-plastered walls.

NOVELTY SHAPES

◄ A STAR-SHAPED WINDOW adds a fanciful touch to this oceanfront child's bedroom. The star motif is carried over into furnishings in the room, including a rug, a pillow, a coverlet, and even the pointed window canopies.

▼ FILLED WITH FREE-FORM LATTICEWORK and a circular opening, this interior window is both a focal point and functional as a pass-through. It also is an interesting way to frame the view from one room into the next.

◄ A CIRCLE-SHAPED WINDOW softens the hard edges of a sharp peak in a gabled wall beneath the ceiling of this bedroom. Centered between two small, horizontal windows, the circular window also adds a sense of symmetry.

Thinking Creatively with Trim

TRIM MOLDING IS INFINITELY VERSATILE. **Even though there's a trim for every purpose, that doesn't mean you can't adapt a piece of chair rail for a concoction of your own design—builders and carpenters have been doing exactly that for centuries. For instance, there's nothing wrong with creating a headboard with elements of a window frame. If you're looking for a bit of trim for a particular application, go to your local builders' supply store and simply play with the shapes. You may find a new baseboard in the crown molding section, or adapt a sill for a picture rail. You also can use almost any material you like, from corrugated metal to driftwood, as long as it's durable enough to stand up to the intended use.**

▲ AN OPEN GRID and a circular window inside a pedimented triangle combine to make an unusual headboard in a sleeping area. Not only is the construction visually arresting, it allows light from the next room to enter the bedroom.

The Rise and Fall of Stairs

A staircase is a complex construction with a substantial presence. Considering that its purpose is to rise from one level of a house to another, it follows that a staircase is composed of many vertical and horizontal elements. There are the treads and risers that make up the steps; the newel post at the foot of the stair; the handrail, or balustrade; the balusters that support the handrail; and a diagonal member, called a stringer, that links everything together. Stairs can rise straight up, turn corners, even curve.

Trim is often integral to specific parts of a staircase. For instance, balusters—the posts between the handrail and the stair steps—can be any number of shapes. handrails are often shaped on the top and have molding-like details on the sides to make them easier to grip. Also, the rhythm of components like steps can be used decoratively. For example, painting the treads (the part you step on) a different color than the risers (the upright part that links each tread to the next) can have a dramatic impact on the look of the stair.

◀ A STAIRWAY CAN ADD DRAMA to your home. Although the decorative elements are kept to a minimum, the sinuous curve of the stair itself and its accompanying handrail combine for maximum impact.

Balusters and Balustrades

Most flights of stairs have both a handrail and balusters—the vertical supports between the stair treads and the handrail. Although balusters come in many styles and shapes, the term usually refers to vertical supports formed by turning on a lathe, a type of machine that shapes circular pieces of wood or stone. Other possibilities for balusters are square or tapered spindles and slats spaced closely together, with or without decorative cutouts. In a formal staircase, turned balusters may have a slightly different pattern, which adds a degree of richness.

The term balustrade means the entire railing, including the top rail and the balusters. Traditional balustrades usually have shaped handrails and either shaped balusters or traditional spindles or slats. For a contemporary look, consider a free-form or asymmetrical balustrade that uses non-traditional baluster patterns or unusual materials, like metal or glass.

◄ THIS TRADITIONAL STAIR, finished with square, matching newel posts at its base, has shaped handrails and spindles for balusters. Smaller versions of the newel posts also appear on the balustrade wherever the stair turns a corner.

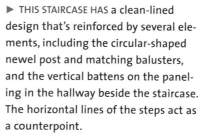

► THIS STAIRCASE HAS a clean-lined design that's reinforced by several elements, including the circular-shaped newel post and matching balusters, and the vertical battens on the paneling in the hallway beside the staircase. The horizontal lines of the steps act as a counterpoint.

◄ AN ELEGANT CURVING BALUSTRADE flows up the stair to the second-floor landing. The balustrade is composed of the handrail, which ends at the bottom of the case in a curving volute, and the balusters, which were made on a lathe (a shaping machine).

▼ MUCH OF THE ELEGANCE of this spiral stair comes from the fact that it is open on one side—a feat for a master builder in times past and present. The graceful, curving balustrade is supported by thin, tapering spindles.

▲ BACKED BY A BANK of floor-to-ceiling windows, a straight spindle stair casts an ever-changing pattern of light on the floor below. A stair rail with white balusters can be particularly effective at keeping an area light and airy.

▲ MANY SPINDLES BEGIN LIFE as long rectangular blocks of wood. They are shaped by a machine called a lathe, which smoothly cuts away extraneous wood. Here, the bottoms of the spindles are square, and the tops are tapered above a bead, a simple decorative element.

Baluster or Banister?

IN A SENSE, the terms baluster and banister are interchangeable: Both refer to vertical supports for railings, as on a stair. If you want to split hairs, however, balusters tend to be more column-like, and are usually cylindrical in shape. They also appear outdoors as well as indoors.

Banisters are usually taller and more slender, like a spindle. But what may be confusing is that banister is also a term that means the handrail on a stair.

► THIS SPINDLE BALUSTRADE is unusual in that it has both top and bottom rails—usually spindles or balusters run directly through to the stair tread. Since much of the stair is made of darkly stained wood, painting the spindles white lightens the effect.

▼ WHEN A STAIRWAY HAS A TURNOUT of only two or three steps at the bottom, the balustrade itself must drop at a right angle. This formal example does so elegantly, finishing in a curving volute.

FREE-FORM BALUSTRADES

▲ THIS UNUSUAL BALUSTRADE is created from individually constructed sections rather than individual or grouped balusters. The tilting lamp shades cleverly play up the asymmetrical lines in the Chinese Chippendale pattern, a design that Thomas Jefferson used on terrace railings at his home, Monticello.

PARTS OF A STAIRCASE

A staircase is a complex assortment of parts, from the treads and risers that make up the steps to the stringer that holds everything together. Some elements, like balusters, are traditionally decorative in their own right; others lend themselves to additional trim.

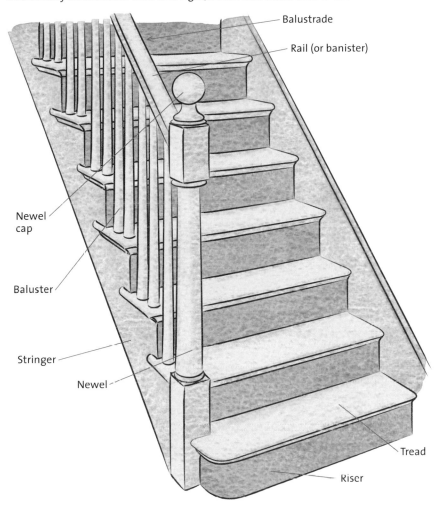

Balustrade

Rail (or banister)

Newel cap

Baluster

Stringer

Newel

Tread

Riser

▲ WHERE MOST TRADITIONAL STAIR RAILS have strong vertical elements in the form of spindles or balusters, this balustrade emphasizes its diagonal lines. The crisp lines of dark wood make a nice counterpoint for the white walls and vertical elements in the room.

◄ NOT EVERY BALUSTRADE is made of wood. This one supports its handrail with cattail-shaped spindles held in place by horizontal wires. Although the effect is contemporary, the strong vertical and horizontal lines are traditional for many types of stairs.

SPINDLES

▶ EXPOSED STAIRS without visible support were unusual in the past, when stairs were typically built from wood. Today, a freestanding stair can be built fairly easily with the help of structural beams. This one is accented with spindles positioned in groups of three.

▲ EVEN A STRAIGHT-AHEAD square spindle creates an interesting pattern in an open stairwell with broad railings on the landings. The regular spacing of two spindles to each stair tread creates its own rhythm. The handrail, painted a dark color, helps to ground the staircase.

◀ SPINDLES ARE CAPABLE of infinite variation. In this curving staircase, spindles in groups of three intersect with a horizontal bar just below the rail. A single spindle that just reaches the bar completes the intricate pattern.

▲ MOST OF THE DECORATIVE IMPACT of this stair with straight spindles and a simple newel post comes from the wood—it came from a deceased elm tree on the property. Embedding the posts into the floor on the second level adds a sense of dimension.

SLATS

▲ A CUTOUT DESIGN in the shape of a falling oak leaf gives a slat staircase and upper balustrade a touch of light-hearted whimsy. Without the peek-a-boo cutouts, the plain white slats might have seemed monotonous.

◄ CUTOUTS ARE A GOOD WAY to lighten the appearance of and add interest to a slatted staircase. Here, two very simple shapes—circles and triangles—alternate in a subtle pattern that gives the natural-wood balustrade extra punch.

Balustrades: Creating Patterns

SLATS MAY BE FLAT and spindles may be straight, but that doesn't mean you can't create inventive patterns in a balustrade that contains either of these simple components. Slats in particular lend themselves to cutout designs in both geometric shapes like circles and squares, and natural shapes such as leaves or flowers. Cutting the shape into the balustrade usually means tracing the pattern on two adjoining slats, then cutting the tracing out with a jigsaw.

Because spindles are three-dimensional, they can be arranged in even more complex patterns. The simplest variations include grouping spindles of the same size in twos or threes. A more elaborate construction might include spindles of different sizes in a repeating pattern, or spindles with cross-bars, crossing patterns, or insets. The possibilities are limited only by your imagination—and the skill of the carpenter building your stair.

▲ THIS DRAMATIC STAIRWAY contains elements that could be adapted to staircases with either spindles or slats. On each step, two narrow spindles flank a wider center spindle. A flat inset with a cutout of a lily spans all three spindles just below the handrail.

◀ THESE COPPER-TILE CUTOUTS are inset into a spindle balustrade that's fairly complex. If you like the idea, you could carry it out more simply by tracing a floral design directly onto closely spaced slats, then cutting them out with a jigsaw.

Newels

A NEWEL CAN BE CONSIDERED THE STARTING POINT for any staircase. It also is typically the finishing point for a handrail. Usually only about 4 ft. high, newels can be square like a post or cylindrical like a column. They also can spiral into a tight curl as a direct extension of the handrail, in a shape called a volute.

Because newels tend to be substantial and prominently placed, it's traditional to decorate them. Square newels are often detailed with recesses, trim molding, and topped with crowns, sometimes with a finial in a three-dimensional shape, such as a carved ball or pineapple. Cylindrical newel posts may display trim similar to that on full-height columns. Contemporary versions of both square and cylindrical shapes can be quite spare, with just a hint of detail.

◄ A RECTANGULAR NEWEL POST has as many sides as a room, so it can support nearly as much trim as four walls. This newel sports a base with concave trim, molding-trimmed recesses, and a beveled cap.

▼ WHEN A STAIR RAILING ENDS in a tightly wound curve, the newel is called a volute. The shape is similar to the tightly curled scrolls on the capital of an Ionic column, which are also called volutes.

◄ NEWEL POSTS SHOULD BE BUILT to comfortable hand height, serving as an anchor for the staircase and a convenient place to put your hand as you begin to ascend the stair. This one is almost within arm's reach of the front door.

▲ A TURNED-OUT STAIR often requires two newels instead of one; it's also customary to repeat the newel design at each stair landing. These newel posts are accented with recessed cutouts and jutting post caps, and topped with round finials. The circular window highlights the finials.

► A NEWEL POST in a Victorian foyer repeats the bull's-eye rosette block commonly found in door and window surrounds of the era, as well as fluting from classical columns. A carved pineapple, a traditional symbol of welcome, makes a fitting cap for the crown.

▲ SINCE NEWELS ARE PATTERNED after columns and posts, it's not surprising that a full-height column looks right at home in place of a newel post in this foyer. An oval window plays off the cylindrical shape of the column.

◄ WHILE NEWEL POSTS with traditional beveled caps and ball finials might seem an odd choice for a contemporary room, they work there because they play off other, subtle traditional details, such as the arches over the doorways.

► THIS SPINDLED STAIRCASE may be short, but it gets a lot of punch from the stout newel post at its foot. Although the post appears solid from a distance, close up you realize it's hollow, thanks to the square cutouts just below the cap.

▼ NEWEL POSTS are seldom cut from solid wood. Instead, they're usually assembled from individual boards shaped into a rectangular post, to which various details are applied. In this example, the square openings give the secret away.

Treads and Risers

THE TREAD IS THE WORKING PART OF THE STAIR—the horizontal supporting step that allows us to put one foot in front of the other and go from one floor to the next. Risers, on the other hand, are the vertical link from one tread to another—at least in traditional staircases. In free-form examples where the treads are invisibly supported, the risers may not exist at all.

Treads and risers often create their own sense of pattern, depending on how the stair is arranged—the rhythmic rise and fall of blocks, the triangular pie shapes of a tightly wound spiral stair. While a tread usually gets little decorative detail on its working surface, you can add some jazz to a staircase by alternating colors (one shade on the treads, the other on the risers). A traditional enhancement is to trim the side of the stair with a decorative scroll called a stair ornament.

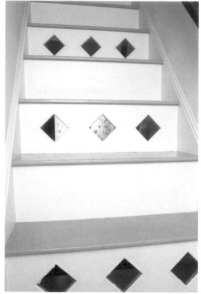

▲ DIAMOND-SHAPED CUTOUTS are an unusual form of decoration on a straight stair with gray treads and white risers. The small openings are a good way to allow both light and air to pass through the staircase into another space.

◄ THE NATURAL ELEGANCE of a spiral staircase is accentuated by the pattern created by the treads and risers, which fan out as they rise. Contrasting the natural wood of the treads with risers painted white adds an extra dimension.

◄ PAINTING BOTH TREADS AND RISERS a matching white creates a clean backdrop for a carpeted stair. Enhancements include a thin band of recessed molding under each tread, and flat, scrolling brackets on the side of each riser.

▼ THIS STAIR GETS A LOT of its visual appeal from the absence of risers and an unusual metal balustrade that falls on the diagonal. Stringers—structural supports on either side of the staircase—hold the horizontal treads in place.

Coloring Treads and Risers

PAINTED STAIRS can lend a dramatic dash of color to a foyer or hall landing. Since these areas seldom have much furniture, your color choices are all but unlimited.

Even a subtle change in color—gray treads over white risers, for example—signals family members or guests where to step. Flashier color combinations accomplish the same goal, but in a bolder way. Since treads take the most punishment, it's usually a good idea to make them darker than the adjacent risers. If you prefer vivid colors, try pairing two shades that are from opposite sides of the color wheel—red and green, for instance. Or try two different colors that have one primary color in common, such as yellow and green. The only rule is that the two colors should look good together. Try holding paint chips for your selections next to each other; the colors should pop in a visually pleasing way.

▲ COLOR ON THE STAIRS plays a big roll in making a transition from a large entry foyer to the upper floors. The yellow-green and light yellow colors chosen for the diamond-patterned floor continue to alternate on the treads and risers.

▶ NOT ALL TREADS AND RISERS are rigid and rectangular. This stair has curving steps that not only complement the oval shape of the room, but gracefully climb and turn to reach a landing only half a story up. The circular window further accents the steps.

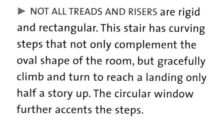

▶ IN AN OPEN STAIRCASE, the up-and-down rhythm of the treads and risers becomes the focal point of the design. Because so many of the other finishes in this room are neutral, the stair is clearly the most important design element in the space.

▲ EVEN WHEN A STAIR IS PAINTED or stained a monochromatic color, subtle decorative details make the difference. The treads are rounded, extend past the flat risers, and are finish-trimmed with concave molding, which gives them added depth—just like a crown molding.

Making Room
for Built-Ins

Built-ins are an ingenious way to create storage, seating, or a focal point for a room where none existed before. For that reason, even the simplest bookcase or seat set into a wall should be viewed with an eye to its decorative possibilities.

Just as the term implies, built-ins are constructions added to interior spaces in a house or apartment. Besides bookcases and window seats, other examples include fireplace surrounds, eating nooks, cabinetry, cubbies, and shelving. While one of the goals of a built-in is usually to make the most of existing space, built-ins can add considerable flair to any room in the house. In addition to its intended purpose, a built-in can establish or enhance a desired decorative theme. A well-designed built-in can also create a sense of focus in almost any place, whether it's a small recess in a wall or an entire room with a vaulted ceiling.

◄ BUILT-INS HELP FINISH A ROOM. Good ones take advantage of views or the presence or absence of sunlight. This corner window seat is probably sunny early in the day and shady in the late afternoon—the perfect time for a nap.

Fireplaces

BUILT-INS LOCATED NEAR A FIREPLACE can be as simple as a single board acting as a mantel, or as elaborate as floor-to-ceiling paneling in wood, carved stone, or cast concrete. Because a fireplace tends to become the focal point of any room, its trim and fittings should reflect a chosen style.

Formal fireplaces are usually composed of elements that are traditionally trimmed. They include the surround (the rectangular area around the fireplace opening), the mantel (the projecting shelf over the fireplace opening), and sometimes, an overmantel (a decorative area between the mantel and ceiling).

It's easy to extend these decorative elements to accommodate built-in bookcases, cabinetry, or even seating. In the Victorian era, it was popular to create seating areas called inglenooks near the hearth. Today, many families add so many built-ins around their fireplaces that the entire area becomes a hearth room, centered around the fireplace.

▲ CLASSICALLY INSPIRED FIREPLACES create their own sense of proportion. Not only are there two of the flat, fluted columns called pilasters on each side, there are also a total of four delicate swags over the fireplace opening— two for each side.

◄ MANTEL TREATMENTS CAN DISGUISE even the plainest fireplace. Although this one is made from concrete block, the mantel, which resembles an upside-down pyramid, makes it look as though it was constructed in the Art Deco era. It's flanked on either side by built-in cabinets.

▶ IN AN EARLY AMERICAN HOME, the surround for a fireplace encompasses an entire wall decorated with moldings and the flat, fluted columns called pilasters. Although the wall appears flush, it was common to conceal a secret cubby or two behind the paneled walls.

TRIM FOR A FIREPLACE

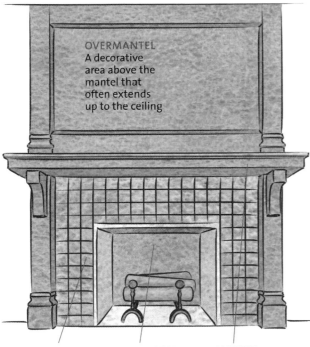

OVERMANTEL
A decorative area above the mantel that often extends up to the ceiling

SURROUND
The rectangular area around the fireplace opening, or firebox; usually trimmed with a non-flammable material

FIREBOX
The fireplace opening

MANTEL
The facing around the fireplace opening and surround, including the projecting shelf above it

◄ THIS MANTELPIECE in the early American Georgian style includes an overmantel, a type of decorative trimwork that extends between the mantel and the ceiling. An overmantel creates a natural frame for a picture or a mirror.

► THE SHALLOW, POINTED ARCH in this mantel is reminiscent of the Tudor style of old England. The shaped overmantel—easy enough for a skilled craftsman to create from drywall or plaster—gracefully conceals the flue and adds more definition to the room.

▲ THE OWNER OF THIS HOUSE did each of the mantels in various rooms in a different style, from French Renaissance to rustic. This one incorporates elaborate carving on the wood overmantel, which is tapered to suggest the shape of the receding flue.

◄ THIS MINIMALIST FIREPLACE faced with concrete has only the vestige of a mantel—yet it serves several purposes. Installed off-center, the narrow mantel makes a handy shelf for displaying art work, while a deep niche in the surround holds firewood.

INFORMAL FIREPLACES

▲ ALTHOUGH THE PROPORTIONS and detail are formal, this mantelpiece gets a rustic feeling from a distressed paint treatment. The shabby-chic look of the paint works well with the rough texture of the floor and adjacent wall.

▶ LITERALLY FILLING A NICHE, a fireplace mantel extends directly into bookcases of matching wood and trim in this contemporary Arts and Crafts interior. Built-ins were an essential part of the Craftsman style—and one reason they are still so popular today.

◄ AN ENTIRE ROOMFUL OF BUILT-INS centers around the fireplace. The cozy space, called an inglenook, is partially enclosed by half-walls. The cushioned seats on either side of the fireplace actually extend out from the hearth.

Mantelpiece Materials

I T'S NATURAL TO THINK OF WOOD as the primary medium for a mantelpiece, but fireplace surrounds have been created from a host of materials over the years, from cast iron to cast concrete.

Because hand-carving has always been costly, the details on many stock and even custom-made wood mantels are actually applied decoration, not carvings. Once the entire piece is stained or painted, these appliqués, or applied decorations, are indistinguishable from the rest of the mantel.

But even the use of appliqués can be expensive. More affordable are mantels created in molds from materials like cast stone, which resembles limestone in appearance. A blend of fine stone and sand combined with a cement-like hardener, the material is very strong. The fact that it's made in a mold allows for fine, crisp detail such as moldings and carvings at the right price.

► WITH ITS FINE DETAILS reminiscent of period moldings, this mantel could be carved from either stone or wood. Instead, it's cast stone, a material molded from fine stone and sand combined with a hardening agent.

Seating

BUILT-IN SEATING FITS IN LITERALLY ANYWHERE. While window seats are a romantic favorite, you can tuck seating into almost any room in the house, either into existing recesses or new ones.

Built-in seating can be utilitarian and functional, as in a small wooden seat in a bathroom that doubles as storage space. It also can be mood-enhancing; a good example would be a long, upholstered seat under a window with a view—the perfect place for relaxing or daydreaming. A good window seat of any size usually serves a dual purpose; the boxy shape is the perfect place to stow toys, blankets, and other bulky items.

Built-in occasional seating works equally well as extra seating in living areas, and site-specific seating in libraries. It's also a favorite for breakfast rooms or dining areas where space is tight: Add a built-in table and you have a breakfast nook.

▼ WHEN THE VIEW INVITES day-dreaming, make the window seat extra roomy. There's enough room for two or even three people in this amply-sized window seat. The handy arm-rail halfway down its length helps keep the cushions in place.

◄ IN A CORNER NEAR an old sash window, this built-in seat makes a cozy spot to while away an afternoon. Paneled with the same darkened beadboard as the walls, it blends beautifully into the room.

► NICHES ARE OFTEN OVERLOOKED as usable space. This one, located beside a bright window, has a pleasing view out toward a patio, making it an ideal location for a built-in seat furnished with a deep cushion and pillows.

Eating Nooks

EVEN THOUGH IT TAKES UP LITTLE SPACE, a breakfast nook can easily become one of the most intimate and beloved spaces in the house. If you'll use it primarily for morning meals, orient the nook toward a window that faces east; if you plan to use at other times, a southern orientation will give you good light all day.

The simplest breakfast nooks usually have a rectangular, fold-down, or fixed table at the center of two narrow benches. Depending on the amount of space you have, you may want to re-arrange the seating into a U- or L-shape, with or without a built-in table. The most comfortable breakfast nooks have cushioned seats built at a height that's versatile enough for both adults and children. Plan your seating area to accommodate the number of people in your household who will share a meal there. The space should feel cozy, not cramped, when all of you sit down together.

▲ THIS BREAKFAST AREA OCCUPIES a corner of a kitchen/living area. The table is fixed, but the benches can be moved in or out, depending on the size of the diners. Plates and cups hang at the ready on the adjacent wall.

◀ A BREAKFAST NOOK surrounded by an assortment of shelves becomes a comfortable corner for eating, watching television, or catching up on reading. The table is portable and can be pulled up to the built-in benches, extended, or pulled completely out of the way.

◄ AN UNADORNED WOODEN BENCH is a good choice for a foyer, where it will take on many roles: a seat for a waiting guest, a place to stack packages, a spot to change out of winter boots, and the occasional guitar performance.

▼ IN A FORMAL DINING ROOM, a built-in bench seat makes a convenient place to sit as guests gather. This one is invitingly cushioned with both upholstered cushions and pillows. It also conceals hidden storage.

▲ SMALL NOOKS AND SPACES that tend to be underused (like dormers) are ideal places for window seats. This one gets extra coziness from its placement below the level of the casement windows and the ample addition of pillows.

▶ A BUILT-IN SEAT gets more use if it's oriented toward light or shade at the time of day it is most likely to be used. This one is set up for either morning breakfast or afternoon tea time.

▶ TURN A HALL or even a foyer into a cozy "away" space for reading or napping with a built-in seat. It also can serve other purposes, such as accommodating a waiting guest or a quick change out of a heavy coat or rain gear.

▲ WE TEND TO THINK of furnishing a seating area with just the right number of sofas and upholstered chairs. This long, low window seat is easily part of the conversation zone, yet it doesn't make the room feel cramped.

▶ A WELL-PLANNED WINDOW SEAT can become part of the architecture of a room, or even its focal point. This one includes several elements key to the room's English Arts and Crafts theme, including diamond-paned windows, crisp white trim moldings, and period-inspired fabric on the cushion.

Cabinets and Shelving

CABINETS AND SHELVING ARE SO ESSENTIAL to a home they've almost become an afterthought. Yet well-planned casework or built-in storage can free up living space and make your house feel much larger. And although built-ins are usually more expensive than freestanding pieces, good built-ins usually pay off in terms of real estate value (or at least, they make it easier to sell your home).

Cabinets and other casework such as built-in dressers are usually enclosed by drawers or doors. Shelves are usually open, although in some cases, they may be enclosed by glass or paneled doors. Cubbies can be in any shape or form, open or closed. Often, the distinctions between the various forms blur as different components combine in one or several built-ins.

◄ A BUILT-IN BUREAU takes up very little space in a bedroom. Because it's recessed into the wall, it doesn't interfere with traffic flow, and items on top of the dresser don't easily disappear behind the back.

▲ BUILT-IN STORAGE doesn't have to be large to be useful. This framed mirror contains two shallow shelves; it's in just the right place to accommodate a reader who wants to curl up with a good book.

▲ IF YOU HAVE A SIZABLE COLLECTION of books or collectibles, a built-in bookcase is a great way to showcase your treasures. This one has a strong vertical pattern at the center that plays up the horizontal pattern of the railing above.

▶ BUILT-INS MAKE THE MOST of a small bedroom in a spectacular location. Not only is there a built-in side table and shelving, even the bed is built in. Additional storage for books is tucked underneath the frame of the bed.

▲ IN A BEDROOM with a vaulted ceiling at oddly shaped angles, built-in armoires fitted into the voids on either side of the fireplace add weight, color, and texture, establishing a sense of style for the room.

▶ A HOUSE WITH LARGE OPEN SPACES can really benefit from built-ins that not only provide needed storage, but also create divisions between spaces. Using the same blond trim wood on the fireplace and cabinetry helps unify the look of a room that's largely open.

◄ BUILT-INS ARE A GREAT WAY to unify the architectural trim in a room. This dresser unifies all the storage in one place, clearing up the potential for clutter, and it literally ties into the crown molding.

► A CUSTOM BUILT-IN can be as complex—and expensive—as a fine piece of furniture, as this bathroom vanity demonstrates. The best built-ins are made by cabinetmakers. The measurements are often so exact that the piece fits precisely into the intended space.

► BUILT-INS HAVE A LONG HISTORY in our country, going back to the corner cupboard. Just as contemporary built-ins make the most of available space, the corner cabinet takes up one of the least-used spaces in a formal room.

SHELVES AND CUBBIES

▼ A DEEP WALL (or one that backs up to another space fitted with built-ins, like a kitchen) is a good spot for a built-in bar. This unit includes a lower cabinet, a glass shelf, and a wine rack.

▶ IN A ROOM WITH FORMAL PROPOR-TIONS, matched bookcases on either side of a fireplace can reinforce the sense of balance. You can get the basic materials for the shelves—including side supports with graduated spacer holes for shelf hardware—at most home centers.

▶ BOOKCASES BUILT just below large windows that wrap around a corner take up very little space in this living room. They also serve two purposes: storage for books below, and a flat surface for lamps, plants, and the occasional piece of artwork above.

▲ IF YOU HAVE A LARGE BOOK COLLECTION, you can create a library for it on a single wall in a space only 12 in. to 15 in. deep. This space, for example, can accommodate more than 700 books.

◄ LARGE OPEN CUBBIES can make a home for artwork, books, or even plants if enough light is available. These are not packed too tightly with items, which helps give the room a clean, open look.

Resources

BOOKS AND MAGAZINES

Among the most helpful publications in preparing this book was *Fine Homebuilding*, a remarkable magazine that manages to straddle the line between historic and contemporary design and traditional and 21st-century building techniques with considerable finesse. You can search *Fine Homebuilding's* online archive for stories of interest at finehomebuilding.com; a recent search using the term "woodwork" produced more than 150 previously published stories. For subscription information, contact The Taunton Press, 63 South Main St., Newtown, CT, 06470; (800) 477-8727.

One of the best sources for locating trim of all kinds is the Design Center at oldhouseinteriors.com, where you can search for products by key word. (For example, type in the word "moulding" — purists prefer the older spelling of the word—and it will lead you to 36 companies that specialize in plaster, wood, or synthetic trim.) *Old-House Interiors*, the magazine of period-inspired home design, also publishes a lavishly illustrated, hold-in-your-hand companion volume, the *Design Center Sourcebook*, each year. Look for trim dealers and products under the categories "Walls & Ceilings" and "House & Garden Details." It's sold on newsstands and direct from Gloucester Publishers, 108 E. Main St., Gloucester, MA 01930; (978) 283-3200.

From the standpoint of learning about trim, the most influential magazine I've ever read is *Old-House Journal*—first as a subscriber and later as a staff editor on the publication for four years, when it was owned and managed by Gloucester Publishers, the parent company of *Old-House Interiors*. Now owned by Restore

Media, the magazine maintains an online website (oldhousejournal.com) and publishes an annual *Restoration Directory*, a good source for makers and dealers of all types of traditional building products, including trim. Old-House Journal's Restoration Directory, 1000 Potomac St. NW, Suite 102, Washington, DC 20007; (202) 339-0744.

There are a number of recently published books that can help you learn more about trim. They include:

Silber, Jay. *Decorating with Architectural Trimwork.* **Upper Saddle River, NJ: Creative Homeowner, 2001.** Although intended for those who want to create their own woodwork, this well-illustrated guide gives many examples of trim suitable for the homes most of us live in. Drawings and close-up photography are especially helpful if you are looking for detailed examples of typical trim parts, like chair rails, wall paneling, and crown moldings.

Wedlick, Dennis. *Good House Parts: Creating a Great Home Piece by Piece.* **Newtown, CT: The Taunton Press, 2003.** Based on the premise that a house is more than the sum of its parts, an architect familiar to readers of Taunton publications like *Fine Homebuilding* explores the possibilities of house parts as small as a newel post to structures that span an entire home, like the roof or the floor.

Dekorne, Clayton. *Trim Carpentry and Built-Ins.* **Newtown, CT: The Taunton Press, 2002.** Part of the "Build Like a Pro" series from The Taunton Press, this book is intended for an audience with some knowledge of construction and building

techniques. Even for the uninitiated, it's still a great way to learn about how trim works in your home, whether it's a simple construction like a window sill or a complex one like a staircase.

FINDING TRIM FOR YOUR HOME

The best place to start looking for trim is your local home center or builder's supply store, where you'll find the basic building blocks for wall trim, from simple crown moldings to cast plaster ceiling ornaments. If you don't see what you're looking for or simply want a more custom look, there are hundreds of specialty retailers who offer both stock and custom trim components.

Think of the list below as more of an overview than an exhaustive source list. At the same time, I've tried to include dealers who offer hard to find specialty products, like hand-carved enrichments or bendable moldings.

Wall and Ceiling Trim

Architectural Products by Outwater (EAST)
4 Passaic St.
Wood-Ridge, NJ 07075
(800) 631-8375
archpro.com

Architectural Products by Outwater (WEST)
4720 W. Van Buren
Phoenix, AZ 85043
(800) 248-2067
Archpro.com

Decorator's Supply Co.
3610 S. Morgan St.
Chicago, IL 60609
(773) 847-6300
decoratorssupply.com

Wood

American Custom Millwork
P.O. Box 3608
Albany, GA 31701
(229) 888-3303
acmi-inc.com

Bendix Mouldings
37 Ramland Rd. South
Orangeburg, NY 10962
(800) 526-0240
bendixmouldings.com

Cumberland Woodcraft
P.O. Box 609
Carlisle, PA 17013
(800) 367-1884
cumberlandwoodcraft.com

Mad River Woodworks
P.O. Box 1067
Blue Lake, CA 95525
(707) 668-5671
madriverwoodworks.com

Pioneer Millworks
1180 Commercial Dr.
Farmington, NY 14425
(800) 951-9663
pioneermillworks.com

Vintage Woodworks
P.O. Box 39
Quinlan, TX 75474
(903) 356-2158
vintagewoodworks.com

White River Hardwoods (also synthetic moldings)
1197 Happy Hollow Rd.
Fayetteville, AR 72701
(800) 558-0119
mouldings.com

Plaster

Classic Mouldings
226 Toryork Dr.
Toronto, ON M9L 1Y1
866) 745-5560
classicmouldings.com

Felber Ornamental Plastering
100 West Washington St.
Norristown, PA 19404
(800) 392-6896
felber.net

Fischer & Jirouch
4821 Superior Ave.
Cleveland, OH 44103
(216) 361-3840
fischerandjirouch.com

Synthetics (polymers and urethanes)

Focal Point Architectural Products
3006 Anaconda Dr.
Tarboro, NC 27886
(866) 843-2880
focalpointap.com

Fypon Ltd.
P.O. Box 248
Seven Valleys, PA 17360
(800) 955-5748
fypon.com

Style Solutions Inc. (also flexible mouldings)
960 West Barre Rd.
Archbold, OH 43502
(800) 446-3040
stylesolutionsinc.com

Enrichments and Appliqués

Enkeboll Designs
16506 Avalon Blvd.
Carson, CA 90746
(800) 745-5507
enkeboll.com

J.P. Weaver
941 Air Way
Glendale, CA 91201
(818) 500-1740
jpweaver.com

Bendable Moldings

B.H. Davis Company
P.O. Box 70
Grosvenordale, CT 06246
(860) 923-2771
curvedmouldings.com

ResinArt East
201 Old Airport Rd.
Fletcher, NC 28732
(800) 497-4376
resinart.com

ResinArt West
1621 Placentia Ave.
Costa Mesa, CA 92627
(800) 258-8820
resinart.com

Flex Moulding, Inc.
16 East Lafayette St.
Hackensack, NJ 07601
(800) 307.3357

Wall Paneling

Architectural Paneling, Inc.
D&D Building
979 Third Ave., Suite 919
New York, NY 10022
(212) 371 9632
apaneling.com

New England Classic Interiors
3 Adams St.
South Portland, ME 04106
(888) 880-6324
newenglandclassic.com

Embossed "Tin" Ceilings

AA Abbingdon Affiliates
2149-51 Utica Ave.
Brooklyn, NY 11234
(718) 258-8333
abbingdon.com

Chelsea Decorative Metal
8212 Braewick Dr.
Houston, Texas 77074
(713) 721-9200
thetinman.com

M-Boss
5350 Grant Ave.
Cleveland, OH 44125
(866) 886-2677
mbossinc.com

Snelling's Thermo-Vac
P.O. Box 210
Blanchard, LA 71009
(318) 929-7398
ceilingsmagnifique.com

W. F. Norman Corp.
214 N. Cedar
Nevada, MO 64772
(800) 641-4038
wfnorman.com

Ceiling Paneling Systems

Global Specialty Products Ltd.
976 Hwy 212 East
Chaska, MN 55318
(800) 964-1186
surfacingsolution.com

Reclaimed Wood/Beams

Conklin's Authentic Antique Barnwood
RR 1, Box 70
Susquehanna, PA 18847
(570) 465-3832
conklinsbarnwood.com

Mountain Lumber Company
P.O. Box 289
Ruckersville, VA 22968
(800) 445 2671
mountainlumber.com

Ramase
661 Washington Rd.
Woodbury, CT 06798
(203) 263-3332 or (800) WIDE-OAK
ramase.com

TerraMai
1104 Firenze St.
McCloud, CA 96057
(800) 220-9062
terramai.com

Vintage Lumber
1 Council Dr.
Woodsboro, MD 21798
Vintagelumber.com
(800) 499-7859

Columns and Capitals

Chadsworth's 1-800-COLUMNS
277 N. Front St.
Wilmington, NC 28401
(800) 265-8667
columns.com

Columns Anywhere
4744 Dovecote Trail
Suwanee, GA 30024
(770) 831-3681
ColumnsAnywhere.com

HB&G
P.O. Box 589
Troy, AL 36081
(800) 264-4424
hbgcolumns.com

Worthington Group Ltd.
P.O. Box 868
Troy, AL 36081
(800) 872-1608
worthingtonmillwork.com

Staircases

Adams Stair Works
1083 S. Corporate Circle
Grayslake, IL 60030
(847) 223 1177
adamsstair.com

Curvoflite
205 Spencer Ave.
Chelsea, MA 02150
(617) 889-0007
curvoflite.com

Stairworld Inc.
2-110 Bentley Ave.
Ottawa, ON K2E 6T9 Canada
(800) 387-7711
stairworld.com

York Spiral Stair
720 Main St.
Vasselboro, ME 04989
(800) 996 5558
yorkspiralstair.com

Fireplace Mantels

DMS Studios
5-50 51st Ave.
Long Island City, NY 11101
(718) 937-5648
dmsstudios.com

Old World Stoneworks
5400 Miller Ave.
Dallas, TX 75206
(800) 600-8336
oldworldstoneworks.com

Stone Magic
301 Pleasant Dr.
Dallas, TX 75217
(800) 597-3606
stonemagic.com

Credits

CHAPTER 1

p. 4: Photo © 2004 Carolyn L. Bates–carolynbates.com; p. 6: (left) Photo © davidduncanlivingston.com; (right) Photo © Brian Vanden Brink, photographer; p. 7: Photo © Tim Street-Porter; p. 8: (left) Photo © Brian Vanden Brink, photographer; (right) Photo © Jonathan Wallen 2004; p. 10: (top) Photo © 2004 Carolyn L. Bates–carolynbates.com; (bottom) Photo © Jonathan Wallen 2004; p. 11: Photo © davidduncanlivingston.com; p. 12 (top) Photo © Scot Zimmerman; (bottom) Photo © davidduncanlivingston.com; p. 13: Photo © Chipper Hatter; p. 14: Photo © davidduncanlivingston.com; p. 15: (left) Photo © Tim Street-Porter; (right) Photo © davidduncanlivingston.com

CHAPTER 2

p. 16: Photo © davidduncanlivingston.com; p. 18: Photo by Andy Engel, © The Taunton Press, Inc.; pp. 18–19: Photo © Grey Crawford; p. 20: (left) Photo © Scot Zimmerman; (right) Photo © Chipper Hatter; p. 21: (top) Photo © Sandy Agrafiotis; (bottom) Photo © Chipper Hatter; p. 22: (top) Photo by Charles Bickford,© The Taunton Press, Inc.; (bottom) Photo © Chipper Hatter; p. 23: (top) Photo © Jonathan Wallen 2004; (bottom) Photo by Roe A. Osborn, © The Taunton Press, Inc.; p. 24: Photo © Sandy Agrafiotis; p. 25: (left) Photo © Brian Vanden Brink, photographer; (right) Photo © Jonathan Wallen 2004; p. 26: (left) Photo © Brian Vanden Brink, photographer; (lower right) Photo © Jonathan Wallen 2004; pp. 26–27: Photo © Scot Zimmerman; p. 27: Photo © Tim Street-Porter; p. 28: (left) Photo © Tim Street-Porter; (right) Photo ©

Jonathan Wallen 2004; p. 29: (top) Photo © Scot Zimmerman; (bottom) Photo © 2004 Carolyn L. Bates–carolynbates.com; p. 30: (left) Photo © Tim Street-Porter; (right) Photo © Grey Crawford; Photo © 2004 Carolyn L. Bates–carolynbates.com; p. 31: Photo © 2004 Carolyn L. Bates–carolynbates.com; p. 32: (top) Photo © Scot Zimmerman; (bottom) Photo © Rob Karosis/www.robkarosis.com; p. 33: Photo © Scot Zimmerman; p. 34: (bottom) Photo © davidduncanlivingston.com; pp. 34–35: Photo © Brian Vanden Brink, photographer; p. 35: Photo © 2004 Carolyn L. Bates–carolynbates.com; p. 36: Photo © Jonathan Wallen 2004; pp. 36–37: Photo © davidduncanlivingston.com; p. 37: Photo © davidduncanlivingston.com; p. 38: (top) Photo © Scot Zimmerman; (bottom) Photo © davidduncanlivingston.com; p. 39: (top) Photo © Sandy Agrafiotis; (bottom) Photo © Brian Vanden Brink, photographer; p. 40: (top) Photo © davidduncanlivingston.com; (bottom) Photo © Jonathan Wallen 2004; p. 41: (top) Photo © 2004 Carolyn L. Bates–carolynbates.com; (bottom) Photo © Sandy Agrafiotis; p. 42: (left) Photo © Chipper Hatter; (right) Photo © Chipper Hatter; pp. 42–43: (top) Photo © davidduncanlivingston.com; p. 44: Photo © davidduncanlivingston.com; p. 45: (left) Photo © Brian Vanden Brink, photographer; (right) Photo © Chipper Hatter; p. 47: (top) Photo © Scot Zimmerman; (bottom) Photo © Brian Vanden Brink, photographer; p. 47: Photo © Brian Vanden Brink, photographer

CHAPTER 3

p. 48: Photo © Sandy Agrafiotis; p. 50: Photo © Chipper Hatter; pp. 50–51: (top) Photo © Sandy

Agrafiotis; (bottom) Photo © 2004 Carolyn L. Bates–carolynbates.com; p. 51: Photo by Charles Miller, © The Taunton Press, Inc.; p. 52: Photo © Chipper Hatter; pp. 52–53: (top) Photo © Chipper Hatter; (bottom) Photo © Jonathan Wallen 2004; p. 53: Photo © davidduncanlivingston.com; p. 54: Photo © davidduncanlivingston.com; p. 55: (top) Photo © Scot Zimmerman; (bottom left) Photo © Scot Zimmerman; (bottom right) Photo © davidduncanlivingston.com; p. 56: Photo © Robert Perron, phtotographer; pp. 56–57: (top) Photo © Chipper Hatter; (bottom) Photo © davidduncanlivingston.com; p. 57: Photo © Scot Zimmerman; p. 58: (left) Photo © davidduncanlivingston.com; (right) Photo courtesy of HB&G Building Projects; p. 59: Photo © Rob Karosis/www.robkarosis.com; p. 60: (left) Photo © Brian Vanden Brink, photographer; (top right) Photo © Jonathan Wallen 2004; (bottom right) Photo © Chipper Hatter; p. 61: (top) Photo © 2004 Carolyn L. Bates–carolynbates.com; (bottom) Photo © Scot Zimmerman; p. 62: Photo © davidduncanlivingston.com; pp. 62–63: Photo © Tim Street-Porter; p. 63: (left) Photo by Charles Bickford, © The Taunton Press, Inc.; (right) Photo © Tim Street-Porter; p. 64: (left) Photo © Tim Street-Porter; (right) Photo © davidduncanlivingston.com; p. 65: (top) Photo © davidduncanlivingston.com; (bottom) Photo by Roe A. Osborn, © The Taunton Press, Inc.; p. 66: (top) Photo © Sandy Agrafiotis; (bottom) Photo © Scot Zimmerman; pp. 66–67: Photo © Tim Street-Porter; p. 67: Photo © Scot Zimmerman; p. 68: (top) Photo by Andy Engel, © The Taunton Press, Inc.; (bottom) Photo © Tim Street-Porter; p. 69: (top)

Photo © 2004 Carolyn L. Bates–carolynbates.com; (bottom) Photo © Tim Street-Porter; p. 70: Photo © Brian Vanden Brink, photographer; pp. 70–71: (top) Photo © Tim Street-Porter; (bottom) Photo © Sandy Agrafiotis; p. 71: Photo by Roe A. Osborn, © The Taunton Press, Inc.; p. 72: (left) Photo © Ken Gutmaker Photography; (top right) Photo © Scot Zimmerman; (bottom right) Photo by Charles Bickford, © The Taunton Press, Inc.; p. 73: (left) Photo © Tim Street-Porter; (right) Photo © davidduncanlivingston.com

CHAPTER 4

p. 74: Photo © Jonathan Wallen 2004; p. 76: (left) Photo © Sandy Agrafiotis; (right) Photo © Brian Vanden Brink, photographer; p. 77 Photo by Roe A. Osborn, © The Taunton Press, Inc.; p. 78 Photo © William P. Wright; pp. 78–79 Photo © Tim Street-Porter; p. 79 Photo © Robert Perron, photographer; p. 80 Photo © Brian Vanden Brink, photographer; pp. 80–81 Photo by Charles Bickford, © The Taunton Press, Inc.; p. 81 (left) Photo © davidduncanlivingston.com; (right) Photo © 2004 Carolyn L. Bates–carolynbates.com; p. 82 (left) Photo © Brian Vanden Brink, photographer; (top right) Photo © Brian Vanden Brink, photographer; (bottom right) Photo © Brian Vanden Brink, photographer; p. 83 Photo © Brian Vanden Brink, photographer; p. 84 Photo © davidduncanlivingston.com; pp. 84–85 Photo © Chipper Hatter; p. 85 Photo © Brian Vanden Brink, photographer; p. 86 (left) Photo © davidduncanlivingston.com; (right) Photo © 2004 Carolyn L. Bates–carolynbates.com; pp. 86–87 Photo © Chipper Hatter; p. 87 Photo © Chipper Hatter; p. 88 (top) Photo © Scot

Zimmerman; (bottom) Photo © 2004 Carolyn L. Bates–carolynbates.com; p. 89 (left) Photo © davidduncanlivingston.com; (right) Photo © Brian Vanden Brink, photographer; p. 90 Photos © Rob Karosis/www.robkarosis.com; p. 91 Photo © davidduncanlivingston.com; p. 92 (top) Photo © davidduncanlivingston.com; (bottom) Photo © Scot Zimmerman; p. 93 (top) Photo © davidduncanlivingston.com; (bottom) Photo © Scot Zimmerman; p. 94 Photo © Brian Vanden Brink, photographer; pp. 94 95 Photo © Scot Zimmerman; p. 95 (top) Photo © Chipper Hatter; (bottom) Photo © Sandy Agrafiotis

CHAPTER 5

p. 96 Photo © Sandy Agrafiotis; p. 98 (top) Photo © Rob Karosis/www.robkarosis.com; (bottom left) Photo © Brian Vanden Brink, photographer; (bottom right) Photo © davidduncanlivingston.com; p. 99 Photo © Chipper Hatter; p. 100 (left) Photo © Chipper Hatter; (right) Photo © davidduncanlivingston.com; pp. 100–101 Photo © Jonathan Wallen 2004; p. 101 Photo © Scot Zimmerman; p. 102 (top left) Photo © Brian Vanden Brink, photographer; (bottom left) Photo © Rob Karosis/www.robkarosis.com; (right) Photo © Brian Vanden Brink, photographer; p. 103 (top) Photo © 2004 Carolyn L. Bates–carolynbates.com; (bottom) Photography © Jason McConathy; p. 104 (left) Photo by Scott Gibson, © The Taunton Press, Inc.; (top right) Photo © Robert Perron, photographer; (bottom right) Photo © Rob Karosis/www.robkarosis.com; p. 105 (top) Photo © Robert Perron, photographer; (bottom) Photo © Scot Zimmerman; p. 106 Photo by Andy Engel, © The Taunton Press, Inc.; p. 107 (top

left) Photo by Todd Caverly, photographer–Brian Vanden Brink photos © 2004; (bottom left) Photo © Sandy Agrafiotis; (right) Photo © Brian Vanden Brink, photographer; p. 108 (left) Photo © Sandy Agrafiotis; (right) Photo © davidduncanlivingston.com ; p. 109 (left) Photo © 2004 Carolyn L. Bates–carolynbates.com; (right) Photo © Brian Vanden Brink, photographer

CHAPTER 6

p. 110 Photo © Brian Vanden Brink, photographer; p. 112 Photo © Brian Vanden Brink, photographer; pp. 112–113 (top) Photo © Scot Zimmerman; (bottom) Photo © Brian Vanden Brink, photographer; p. 113 Photo © 2004 Carolyn L. Bates–carolynbates.com; p. 114 (left) Photo © Robert Perron, photographer; (right) Photo by Scott Gibson, © The Taunton Press, Inc.; p. 115 (top) Photo © Robert Perron, photographer; (bottom) Photo © Sandy Agrafiotis; p. 116 Photo © Chipper Hatter; p. 117 Photos © davidduncanlivingston.com; p. 118 (top) Photo © Scot Zimmerman; (bottom left) Photo © Brian Vanden Brink, photographer, (bottom right) Photo © Scot Zimmerman; p. 119 Photo by Roe A. Osborn, © The Taunton Press, Inc.; p. 120 (left) Photo by Charles Bickford, © The Taunton Press, Inc.; (right) Photo © Tim Street-Porter; p. 121 Photos by Charles Miller, © The Taunton Press, Inc.; p. 122 Photo © Chipper Hatter; pp. 122–123 Photo © Brian Vanden Brink, photographer; p. 123 (top) Photo © Grey Crawford; (bottom) Photo © davidduncanlivingston.com; p. 124 (top left) Photo © Scot Zimmerman; (bottom left) Photo © davidduncanlivingston.com; (right) Photo by Chris Green, © The Taunton Press, Inc.; p. 125 Photos by Andy Engel, © The Taunton Press, Inc.; p. 126 (left)

Photo © Chipper Hatter; (right) Photo © Robert Perron, photographer; p. 127 (left) Photo © Brian Vanden Brink, photographer; (right) Photo © Jonathan Wallen 2004; p. 128 Photo © Robert Perron, photographer; pp. 128–129 Photo © Chipper Hatter; p. 129 (top) Photo © Tim Street-Porter; (bottom) Photo © 2004 Carolyn L. Bates–carolynbates.com

CHAPTER 7

p. 130 Photo © davidduncanlivingston.com; p. 132 (bottom) Photo by Scott Gibson, © The Taunton Press, Inc.; (top) Photo © Tim Street-Porter; p. 133 Photo © Sandy Agrafiotis; p. 134 (left) Photo © 2004 Carolyn L. Bates–carolynbates.com; (right) Photo © davidduncanlivingston.com; p. 135 (top) Photo © Tim Street-Porter; (bottom) Photo © Tim Street-Porter; p. 136 (top) Photo © Tim Street-Porter; (bottom) Photo © davidduncanlivingston.com; p. 137 (top) Photo © Grey Crawford; (bottom) Cast Stone Mantel photo courtesy of Old World Stoneworks; p. 138 Photo © Brian Vanden Brink, photographer; p. 139 (top) Photo © Jonathan Wallen 2004; (bottom) Photo © davidduncanlivingston.com; p. 140 (top) Photo © davidduncanlivingston.com; (bottom) Photo © Sandy Agrafiotis; p. 141 (top) Photo © Brian Vanden Brink, photographer; (bottom) Photo © davidduncanlivingston.com; (left) Photo by Charles Bickford, © The Taunton Press, Inc.; (bottom right) Photo © Tim Street-Porter; p. 143 (top right) Photo © Brian Vanden Brink, photographer; (top left) Photo © davidduncanlivingston.com; (bottom right) Photo © 2004 Carolyn L. Bates–carolynbates.com; p. 144 Photo © Brian Vanden Brink, photographer; p. 145 (top and bottom right) Photo © davidduncanliv-

ingston.com; (top left) Photo © Jonathan Wallen 2004; p. 146 (left) Photo by Chris Green, © The Taunton Press, Inc.; (bottom and top right) Photo © davidduncanlivingston.com; p. 147 (top) Photo © Chipper Hatter; (bottom) Photo © Sandy Agrafiotis; p. 148 (left and bottom right) Photo © davidduncanlivingston.com; (top right) Photo © Scot Zimmerman; p. 149 Photos © davidduncanlivingston.com.

For More Great Design Ideas, Look for These and Other Taunton Press Books wherever Books are Sold.

NEW KITCHEN IDEA BOOK
ISBN 1-56158-693-5
Product #070773
$19.95 U.S.
$27.95 Canada

NEW BATHROOM IDEA BOOK
ISBN 1-56158-692-7
Product #070774
$19.95 U.S.
$27.95 Canada

NEW KIDSPACE IDEA BOOK
ISBN 1-56158-694-3
Product #070776
$19.95 U.S.
$27.95 Canada

TAUNTON'S HOME STORAGE IDEA BOOK
ISBN 1-56158-676-5
Product #070758
$19.95 U.S.
$27.95 Canada

NEW BUILT-INS IDEA BOOK
ISBN 1-56158-673-0
Product #070755
$19.95 U.S.
$27.95 Canada

TILE IDEA BOOK
ISBN 1-56158-709-5
Product #070785
$19.95 U.S.
$27.95 Canada

TAUNTON'S FAMILY HOME IDEA BOOK
ISBN 1-56158-729-X
Product #070789
$19.95 U.S.
$27.95 Canada

TAUNTON'S HOME WORKSPACE IDEA BOOK
ISBN 1-56158-701-X
Product #070783
$19.95 U.S.
$27.95 Canada

BACKYARD IDEA BOOK
ISBN 1-56158-667-6
Product #070749
$19.95 U.S.
$27.95 Canada

POOL IDEA BOOK
ISBN 1-56158-764-8
Product #070825
$19.95 U.S.
$27.95 Canada

DECK & PATIO IDEA BOOK
ISBN 1-56158-639-0
Product #070718
$19.95 U.S.
$27.95 Canada

TAUNTON'S FRONT YARD IDEA BOOK
ISBN 1-56158-519-X
Product #070621
$19.95 U.S.
$27.95 Canada